# Robust Flight Control
# Several Aeronautical Applications



**Samir Bennani**

# Foreword

The list of persons involved will certainly not be complete but in first instance I would like to express my gratitude to the initiator of this enterprise, who came up with the idea to let me conduct this research, Ton van der Weiden.

This research has been made possible thanks to my promotor prof. Mulder, prof. Bosgra who had the faith in the possibilities and opportunities for the application of robust control to flight control systems.

My interest for robust control has only seriously started in 1990 where I took the $\mu$-short course on theory and applications of robust multivariable control by Doyle, Packard, Balas and Glover. Since then, a fruitful communication and cooperation has taken place over the years.

I have to admit that it was only since I met Carsten Scherer, that the mathematical essence of robust control became transparent to me. I was always amazed by his talent to make complex things look so simple. I owe him the background necessary to support my design intuition.

I am also thankful to Lieven Vandenberghe for teaching me convex optimization in Leuven, Pascal Gahinet and Pierre Apkarian are endepted for their preliminary versions of LMI-LAB.

I always enjoyed the company of Paul Lambrechts, a friend with whom a long relation led to a successful cooperation in the Garteur FM-AG08 group and many unforgettable congress meeting trips over the world.
I have to mention Jan Terlouw who involved me within the Garteur research group, he has always supported my work at the NLR. The active Garteur participation and experience has brought together so many valuable contacts allowing at an international level contuinity in this exciting research field.

NLR is endepted for the financial support during the time of this research, NWO and SIMONA for the financial funding of our HP machines and Fokker Space & Systems, in particular Piet Lammerste and Ron Huijser for their support during the redaction of the thesis.

Especially, thanks to all my (ex-)students, Gertjan Looye, Dehlia Willemsen, Edwin Nijo, Remco van der Sluis, Corne van Etten, Rogier van der Weerd, and Marieke Siersma with whom I enjoyed the tight cooperation.

Colleagues, Max Mulder, Jan Willem van Staveren, Paul van Gool, Sanjoo Advani, Henk Lindenburg and Peter Kraan are all endepted, we had a fruitful cooperation and enjoyable moments.

Also, my gratitude goes out to Hans van der Vaart for his commnents and review of the original draft of this work. Further I wish to thank all people within and outside the university having supported me.

Finally, this thesis is dedicated to my wife Coralies and family.

# Contents

# List of Figures

# List of Tables

# Notation

### Fields of Numbers

| | |
|---|---|
| $\mathbb{C}$ | Complex numbers |
| $\mathbb{C}_+$ | Open left-half plane |
| $\mathbb{C}_-$ | Open right-half plane |
| $\mathbb{C}^{n}$ | Column vectors of $n$ complex entries |
| $\mathbb{C}^{n\times m}$ | Complex matrices of dimension $n \times m$ |
| $\mathbb{R}$ | Real Numbers |
| $\mathbb{R}^n$ | Column vectors of $n$ real entries |
| $\mathbb{R}^{n\times m}$ | Real matrices of dimension $n \times m$ |

### Sets

| | |
|---|---|
| $\mathbf{B\Delta}$ | Set of Bounded Perturbations |
| $\mathbf{D}$ | Set of Commuting Scalings |
| $\mathbf{D}_{pert}$ | Augmented Set of Commuting Scalings |
| $\mathbf{Q}$ | Set of Unitary Scalings |
| $H_2$ | Subspace of functions in $L_2$ that are analytic in $\mathbb{C}_+$ and uniformingly square integrable for all $\alpha \in \mathbb{R}_+$ |
| $H_\infty$ | Subspace of functions in $L_\infty$ that are analytic and bounded in $\mathbb{C}_+$ |
| $L_2$ | Space of square integrable functions on $j\mathbb{R}$ including $\infty$ |
| $L_\infty$ | Space of bounded functions on $j\mathbb{R}$ including $\infty$ |
| prefix $\mathbb{R}$ | To denote subspace of real-rational functions $\mathbb{R}\mathrm{H}_\infty$. |
| $\mathcal{S}_{j\omega}$ | Set of Complex valued signals |
| $\mathcal{S}_u$ | Set of input signals $u(t)$ in time $t$ |
| $\mathcal{S}_y$ | Set of output signals $y(t)$ in time $t$ |

**Matrices, Operators, Vectors**

| | |
|---|---|
| $diag(A, B, \ldots, Z)$ | $= \begin{bmatrix} A & & & \\ & B & 0 & \\ & 0 & \ddots & \\ & & & Z \end{bmatrix}$ |
| $\begin{bmatrix} A & * \\ B & C \end{bmatrix}$ | $= \begin{bmatrix} A & B^T \\ B & C \end{bmatrix}$ |
| $\begin{bmatrix} A & B \\ C & D \end{bmatrix}$ | State-Space Operator |
| $\left[\begin{array}{c\|c} A & B \\ \hline C & D \end{array}\right]$ | State-Space Realization $G(s) = D + C(sI - A)^{-1}B$ |
| $\begin{bmatrix} M_{11} & M_{21} \\ M_{21} & M_{22} \end{bmatrix}$ | General Analysis System Matrix $M$ |
| $\begin{bmatrix} P_{11} & P_{12} & P_{13} \\ P_{21} & P_{22} & P_{23} \\ P_{31} & P_{32} & P_{33} \end{bmatrix}$ | General Open Loop System Matrix $P$ |
| $Q + AXB + tr.$ | $= Q + AXB + (AXB)^T$ |
| $A + .^T$ | Denotes the Symmetric Matrix $= A + A^T$ |
| $A + .^*$ | Denotes the Hermitian Matrix $= A + A^*$ |
| $A$ | State matrix of the state equation |
| $B$ | Control matrix of the state equation |
| $C$ | State matrix of the output equation |
| $D$ | Direct feedthrough matrix or Scaling Matrix |
| $O$ | Zero matrix |
| $I$ | Unit matrix |
| $I_n$ | Unit matrix of dimension n |
| $K$ | Compensator |
| $P$ | Open Loop Interconnection Matrix |
| $Q$ | Scaling Matrix |
| $R$ | Scaling Matrix |
| $S$ | Scaling Matrix or Sensitivity function |
| $U$ | Unitary Scaling Matrix |
| $V$ | Unitary Scaling Matrix |
| $X$ | Solution to Matrix Inequality |
| $Y$ | Solution to Matrix Inequality |
| $[X]_{ij}$ *or* $x_{ij}$ | The $i, j$ element of matrix |
| $\Delta$ | Perturbation |

| | |
|---|---|
| $\Delta_F$ | Full Perturbation |
| $\Delta_S$ | Scalar Perturbation |
| $j$ | Square root of -1 |
| $\bar{x}$ | Complex conjugate |
| $e$ | Error signal |
| $d$ | Disturbance signal |
| $n$ | Noise signal |
| $r$ | Reference signal |
| $u$ | Control signal |
| $y$ | Measured signal |
| $v$ | Exogenous signal |
| $w$ | Perturbation input signal |
| $z$ | Perturbation output signal |
| $x$ | State vector |
| $\dot{x}$ | Time derivative of the state vector |
| $x^H$ | Tranpose of complex conjugate |
| $x^T$ | Transpose |
| $x^\dagger$ | Pseudo-inverse |
| $\delta$ | Uncertainty element |
| $z_i$ | Zeros of a transfer function |
| $p_i$ | Poles of a transfer function |
| $s$ | Laplace operator |
| $M_p$ | Relative overshoot |
| $t$ | Time |
| $t_r$ | Rise-time |
| $t_s$ | Settling-time |
| $\gamma$ | $H_\infty$-norm or $L_2$-induced norm |
| $\omega_c$ | Cross-over frequency |
| $\omega_b$ | Bandwidth ($-3$ dB) |
| $\omega_n$ | Undamped natural frequency |
| $\omega$ | Circular frequency |
| $\Omega$ | Spatial frequency |
| $\zeta$ | Damping ratio |
| $\tau$ | Time-constant |

### Operation on Matrices

| | |
|---|---|
| $A^T$ | Tranpose of matrix $A$ |
| $A^H$ | Complex conjugate transpose of matrix $A$ |
| $A^{-1}$ | Inverse of matrix $A$ |
| $A^\dagger$ | Pseudo-inverse of matrix $A$ |
| $A^{-T}$ | Inverse transpose or transpose inverse of matrix $A$ |
| $A^{-H}$ | Inverse conjugate transpose or congugate transpose inverse of matrix $A$ |
| $A > 0$ | Hermitian matrix $A = A^H$ with strictly positive eigenvalues |
| $A \geq 0$ | Hermitian matrix $A = A^H$ with non-negative eigenvalues |

| | |
|---|---|
| $A < 0$ | Hermitian matrix $A = A^H$ with strictly negative eigenvalue |
| $A \leq 0$ | Hermitian matrix $A = A^H$ with non-positive eigenvalues |
| $A > B$ | means that $(A - B) < 0$ |
| $adj(A)$ | Adjugate of matrix $A$ |
| $det(A)$ | Determinant of matrix $A$ |
| $rank(A)$ | Rank of matrix $A$ |
| $tr(A)$ | Trace of matrix $A$ |
| $\lambda_i(A)$ | The $i$-th eigenvalue of matrix $A$ |
| $\lambda_{max}(A)$ | Largest Eigenvalue of matrix $A$ |
| $\lambda_{min}(A)$ | Smallest Eigenvalue of matrix $A$ |
| $\lambda_p(A)$ | Perron-Frobenius Eigenvalue of matrix $A$ |
| $\mu_\Delta(A)$ | Structured Singular Value of matrix $A$ with respect to the uncertainty structure $\Delta$ |
| $\rho(A)$ | Spectral Radius of matrix $A$ |
| $\sigma_i(A)$ | The $i$-th Singular Value of matrix $A$ |
| $\bar{\sigma}(A)$ | Largest Singular Value of matrix $A$ |
| $\underline{\sigma}(A)$ | Smallest Singular Value of matrix $A$ |
| $\kappa(A)$ | Condition Number of matrix $A$ |

**Norms on vectors and operators**

| | |
|---|---|
| $\|x\|$ | Modulus of $x$ |
| $\|\|x\|\|_1$ | 1-norm, $\max_j \Sigma_i \|x_{ij}\|$ |
| $\|\|x\|\|_F$ | Frobenius norm |
| $\|\|x\|\|_S$ | Spectral norm |
| $\|\|G\|\|_2$ | $H_2$-norm |
| $\|\|G\|\|_H$ | Hankel norm |
| $\|\|G\|\|_\infty$ | Infinity norm |
| $\|\|G\|\|_\mu$ | $\mu$-"norm" |
| max | Maximum |
| min | Minimum |
| sup | Supremum |
| inf | Infimum |

**Operation on Systems**

| | |
|---|---|
| $G^T$ | Tranpose of the real-rational system $G$, $G^T(s) = G(s)^T$ |
| $G^\sim$ | Adjoint of the real-rational system $G$, $G^\sim = G^T(-s) = G(-$ |
| $G(j\omega)^H$ | Complex conjugate transpose of the frequency response $G(j$ $G(j\omega)^H = G(-j\omega)^T$ |
| $G^{-1}$ | Inverse of the real-rational system $G$, $G^{-1}(s) = G(s)^{-1}$ |
| $G^{-T}$ | Denotes $(G^{-1})^T = (G^T)^{-1}$ |
| $F_l(P, K)$ | Lower Linear Fractional Transformation of $P$ with $K$ |
| $F_u(M, \Delta)$ | Upper Linear Fractional Transformation $M$ with $\Delta$ |

### Transfer Functions

| | |
|---|---|
| $D(s)$ | Transfer function of $D$-scale |
| $D^{-1}(s)$ | Inverse transfer function of $D$-scale |
| $G(s)$ | Transfer function of plant |
| $G_0(s)$ | Nominal Transfer function of plant |
| $\tilde{G}(s)$ | Perturbed transfer function of plant |
| $H(s)$ | Transfer function of plant |
| $K(s)$ | Transfer function of compensator |
| $L(s)$ | Loop Gain transfer function |
| $M(s)$ | Transfer function of analysis plant |
| $P(s)$ | Transfer function of open-loop interconnection structure |
| $\tilde{P}(s)$ | Perturbed open-loop interconnection structure |
| $P_a(s)$ | Augmented open-loop interconnection structure |
| $S(s)$ | Sensitivity function |
| $S_i(s)$ | Input Sensitivity function |
| $S_o(s)$ | Output Sensitivity function |
| $T(s)$ | Complementatry Sensitivity function |
| $T_i(s)$ | Input Complementatry Sensitivity function |
| $T_o(s)$ | Output Complementatry Sensitivity function |
| $W(s)$ | Weighting function |
| $\Delta(s)$ | Complex uncertainty operator |
| $\Delta_a(s)$ | Additive uncertainty operator |
| $\Delta_i(s)$ | Multiplicative uncertainty operator |

### Acronyms Control Theory

| | |
|---|---|
| BRL | Bounded Real Lemma |
| BW | Bandwith |
| dB | decibel |
| DC | Direct Current |
| GM | Gain Margin |
| GS | Gain Scheduling |
| IQC | Integral Quadratic Constraint |
| LMI | Linear Matrix Inequality |
| LFT | Linear Fractional Transformation |
| LQG | Linear Quadratic Gaussian |
| LPV | Linear Parameter Varying |
| LTI | Linear Time Invariant |
| LTV | Linear Time Variant |
| MIMO | Multiple Input Multiple Output |
| NP | Nominal Performance |
| PM | Phase Margin |
| PSD | Power Spectral Density |
| RMS | Root Mean Square |
| RP | Robust Performance |

| | |
|---|---|
| RS | Robust Stability |
| SISO | Single Input Single Output |
| SV | Singular Value |
| SSV | Structured Singular Value |

**Acronyms Flight Control**

| | |
|---|---|
| A/C | Aircraft |
| ACT | Active Control Technology |
| ACTIVE | Advanced Control Technology for Integrated Vehicles |
| AFC | Active Flight Control |
| AFTI | Advanced Fighter Technology Integration |
| AS | Augmentation System |
| AP | Auto Pilot |
| CAS | Control Augmentation System |
| CCV | Computer Configured Vehicle |
| CWS | Control Wheel Steering |
| EAP | Experimental Aircraft Program |
| EFA | European Fighter Aircraft |
| FBW | Fly-By-Wire |
| FC | Flight Control |
| FCS | Flight Control System |
| FTC | Fault Tolerant Control |
| HARV | High Angle of Attack Research Vehicle |
| HIRM | High Incidence Research Model |
| HQ | Handling Qualities |
| ICE | Innovative Control Effectors |
| ILAF | Identical Location for Accelerometer and Force |
| PIO | Pilot Induced Oscillations |
| PRCS | Pitch Rate Command System |
| SCAS | Superaugmented Control Augmentation System |
| SMC | Structural Mode Control |
| MLC | Maneuvre Load Control |
| PCA | Power Controlled Aircraft |
| RESTORE | Reconfigurable Control for Tailless Fighter Aircaft |
| RCAM | Research Civil Aircraft Model |
| RC | Ride Control |
| TAFA | Tailless Advanced Fighter Aircraft |
| TECS | Total Energy Control System |
| THCS | Total Heading Control System |
| VISTA | Variable Incidence System Technology Aircraft |
| VSTOL | Vertical Short Take-Off and Landing |
| WG | Working Group |

**Acronyms Organizations**

| | |
|---|---|
| AGARD | Advisory Group for Aerospace Research and Development |
| DUT | Delft University of Technology |
| GARTEUR | Group for Aeronautical Research and Technology in Europe |
| NASA | National Aerospace & Space Agengy |
| NATO | North Atlantic Treaty Organization |
| NLR | National Aerospace Laboratory |

| Symbol | Alphanumeric | | | Name | Unit |
|---|---|---|---|---|---|
| $p$ | P | x(1) | = | roll rate (in $F_B$) | *rad/s* |
| $q$ | Q | x(2) | = | pitch rate (in $F_B$) | *rad/s* |
| $r$ | R | x(3) | = | yaw rate (in $F_B$) | *rad/s* |
| $\phi$ | PHI | x(4) | = | roll angle (Euler angle) | *rad* |
| $\theta$ | THETA | x(5) | = | pitch angle (Euler angle) | *rad* |
| $\psi$ | PSI | x(6) | = | heading angle (Euler angle) | *rad* |
| $u_B$ | UB | x(7) | = | $x$ component of inertial velocity in $F_B$ | *m/s* |
| $v_B$ | VB | x(8) | = | $y$ component of inertial velocity in $F_B$ | *m/s* |
| $w_B$ | WB | x(9) | = | $z$ component of inertial velocity in $F_B$ | *m/s* |
| $x$ | X | x(10) | = | $x$ position of aircraft CoG in $F_E$ | *m* |
| $y$ | Y | x(11) | = | $y$ position of aircraft CoG in $F_E$ | *m* |
| $z$ | Z | x(12) | = | $z$ position of aircraft CoG in $F_E$ | *m* |

Table 1: Definition RCAM states

| Symbol | Alphanumeric | | | Name | Unit |
|---|---|---|---|---|---|
| $\delta_A$ | DA | u(1) | = | aileron deflection | *rad* |
| $\delta_T$ | DT | u(2) | = | tailplane deflection | *rad* |
| $\delta_R$ | DR | u(3) | = | rudder deflection | *rad* |
| $\delta_{TH_1}$ | THROTTLE1 | u(4) | = | throttle position of engine 1 | *rad* |
| $\delta_{TH_2}$ | THROTTLE2 | u(5) | = | throttle position of engine 2 | *rad* |
| $Wx_E$ | WXE | u(6) | = | Wind velocity in the $x$-axis of $F_E$ | *m/s* |
| $Wy_E$ | WYE | u(7) | = | Wind velocity in the $y$-axis of $F_E$ | *m/s* |
| $Wz_E$ | WZE | u(8) | = | Wind velocity in the $z$-axis of $F_E$ | *m/s* |
| $Wx_B$ | WXB | u(9) | = | Wind velocity in the $x$-axis of $F_B$ | *m/s* |
| $Wy_B$ | WYB | u(10) | = | Wind velocity in the $y$-axis of $F_B$ | *m/s* |
| $Wz_B$ | WZB | u(11) | = | Wind velocity in the $z$-axis of $F_B$ | *m/s* |

Table 2: Definition RCAM inputs

| Symbol | Alphanumeric | | | Name | Unit |
|---|---|---|---|---|---|
| Measured | | | | | |
| $q$ | Q | y(1) | = | pitch rate (in $F_B$) = x(2) | $rad/s$ |
| $n_x$ | NX | y(2) | = | horizontal load factor (in $F_B$) = $\frac{F_x}{mg}$ | - |
| $n_z$ | NZ | y(3) | = | vertical load factor (in $F_B$) = $\frac{F_z}{mg}$ | - |
| $w_V$ | WV | y(4) | = | $z$ component of inertial velocity in $F_V$ | $m/s$ |
| $z$ | Z | y(5) | = | $z$ position of aircraft CoG in $F_E$ = x(12) | $m$ |
| $V_A$ | VA | y(6) | = | air speed | $m/s$ |
| $V$ | V | y(7) | = | total inertial velocity | $m/s$ |
| $\beta$ | BETA | y(8) | = | angle of sideslip | $rad$ |
| $p$ | P | y(9) | = | roll rate (in $F_B$) = x(1) | $rad/s$ |
| $r$ | R | y(10) | = | yaw rate (in $F_B$) = x(3) | $rad/s$ |
| $\phi$ | PHI | y(11) | = | roll angle (Euler angle) = x(4) | $rad$ |
| $u_V$ | UV | y(12) | = | $x$ component of inertial velocity in $F_V$ | $m/s$ |
| $v_V$ | VV | y(13) | = | $y$ component of inertial velocity in $F_V$ | $m/s$ |
| $y$ | Y | y(14) | = | $y$ position of aircraft CoG in $F_E$ = x(11) | $m$ |
| $\chi$ | CHI | y(15) | = | inertial track angle | $rad$ |
| Simulation | | | | | |
| $\psi$ | PSI | y(16) | = | heading angle (Euler angle) = x(6) | $rad$ |
| $\theta$ | THETA | y(17) | = | pitch angle (Euler angle) = x(5) | $rad$ |
| $\alpha$ | ALPHA | y(18) | = | angle of attack | $rad$ |
| $\gamma$ | GAMMA | y(19) | = | inertial flight path angle | $rad$ |
| $x$ | X | y(20) | = | $x$ position of aircraft CoG in $F_E$ = x(10) | $m$ |
| $n_y$ | NY | y(21) | = | lateral load factor (in $F_B$)= $\frac{F_y}{mg}$ | - |

Table 3: Definition RCAM measurements and simulation outputs

| Symbol | | Name | Unit |
|---|---|---|---|
| $\alpha$ | | angle-of-attack | *rad* |
| $q$ | | pitch rate | *rad/s* |
| $n_z$ | | normal acceleration per g | - |
| $\delta$ | | tail fin deflection | *rad* |
| $F_z$ | $C_n(\alpha, \delta, M)0.7p_0M^2S$ | force in $z-$direction | *lbs* |
| $M_y$ | $C_m(\alpha, \delta, M)0.7p_0M^2Sd$ | pitching moment | *ft - lbs* |
| $p_0$ | 973.3 | static pressure at 20,000 ft | $(lb/ft)^2$ |
| $S$ | 0.44 | reference area | $(ft)^2$ |
| $d$ | 0.75 | diameter | $ft$ |
| $m$ | 13.98 | mass of missile | $slugs$ |
| $u$ | $V\cos(\alpha)$ | speed along missile center line | $ft/s$ |
| $V$ | $M \times ss$ | velocity of the missile | $ft/s$ |
| $M$ | $2-4$ | Mach number of the missile | - |
| $ss$ | 1036.4 | speed of sound at 20,000 ft | $ft/s$ |
| $I_y$ | 182.5 | pitch moment of inertia | $(slug-ft)^2$ |
| $g$ | 32.2 | acceleration due to gravity | $(ft/s)^2$ |

Table 4: Missile paramters

# Abstract

This thesis treats the application of a variety of modern robust control techniques as developed in the last decade within the $H_\infty$ framework to the control of the rigid body motion of aircraft.
The general definitions related to $H_\infty$-optimal control and $\mu$-synthesis are first reviewed. In particular a general framework for controller design and analysis consisting of certain matrix measures along with a general system representation called Linear Fractional Transformation (LFT) are shortly reviewed in order to define the notions of Robust Stability, Nominal Performance and Robust Performance.

From a control designer's perspective particular attention was devoted in revealing how the developed machinery can be applied to a two-degrees-of freedom pitch-rate control system design meeting robustly handling qualities requirements.
It is first shown how to incorporate simultaneous performance requirements such as, reference signal tracking, disturbance rejection and control activity limitation through a suitable mathematical formulation in terms of weighting function selection. Rules for weighting function selection to enforce appropriate control solutions trading-off several performance objectives have been established. Having provided design guidelines for nominal performance synthesis the previous design is accomodated to ensure a variety of robustness levels. Using the Structured Singular Value $\mu$ as an indicator for the system's robust performance level it is illustrated how the trade-off can be established between stability robustness requirements and performance requirements.

In a second step, to demonstrate the flexibility of the Structured Singular Value concept to complex multivariable systems common to industrial flight control design problems a benchmark problem has been developed for the Research Civil Aircraft Model (RCAM).
It is shown how to design robustly a control system that achieves multivariable tracking specifications, safety limitations such as stall or roll-angle protection, control coordination such as coordinated turns and smooth altitude or speed captures.
Special emphasis has been placed on demonstrating how to design with guaranteed performance robustness an autopilot behavior similar to the Total Energy Concept System (TECS). The advantage of the robust approach shown here above TECS is that it directly accounts for a more general set of performance requirements while simultaneoulsy handling multiple robustness issues. On the other hand since the TECS concepts relies on basic physical principles deduced from flight mechanics, the TECS controller structure is transparent and easy to schedule with respect to varying operating conditions as opposed to robust control solutions. Nevertheless, the proposed scheduling procedure is system specific and therefore not generally applicable. Also no formal stability guarantees are involved that take the time-varying character of the system into account.

To overcome these limitations, the gain scheduling problem has been addressed formally in the second part of the thesis. The gain scheduling problem formulation relies on the machinery developed in the LTI robust control framework specialized to Linear Parametrically Varying (LPV) systems. Both, the analysis and synthesis results are reformulated as Linear Matrix Inequalities (LMI) for which solutions are obtained using numerical methods from convex optimization.

Integral Quadratic Constraints (IQC) are used to develop scalings for uncertain parameter dependent systems. Special controller parametrization formulae are used to handle a large class of LPV synthesis methods. We have studied two LPV-LFT gain scheduling problems with unbounded paramter rates and one LPV gain scheduling problem with bounded parameter rates. These methods are illustrated and compared on a missile benchmark problem and a Handling Quality (HQ) design for the Cessna Citation that meets Gibson's Control Anticipation Parameter (CAP).

In the final chapter the contributions to flight control system design community is described. Also, conclusions from this work are drawn and recommendations towards future research are given.

Through this work we have experienced that the robust and LPV concepts are particularly well suited in dealing with current and future flight control system design tasks. They are in no way meant to replace classical control but rather than that they are developed to generalize and formalize single-loop concepts to multivariable systems.

# Chapter 1

# Introduction: Flight & Control

*This thesis treats the application of robust control to the design of Flight Control Systems (FCS). The field of application is referred to as Active Flight Control, (AFC). First, some of the possibilities, problems and needs encountered in AFC technology are shortly reviewed.*
*The majority of industrial AFC systems have been designed using classical design tools. We refer to the frequency domain techniques such as Bode, Nyquist and Nichols. These techniques have been successful in tackling most of the Linear Time Invariant (LTI) Single Input Single Output (SISO) aeronautical problems. Difficulties in design are encountered when dealing with more complex problems. This is the case when the problem is multivariable, nonlinear and uncertain on which many performance requirements are placed at the limits of the system capabilities. Further, economic constraints impose the delivery of faster and better designed systems at shrinking development costs. The potentials of multivariable robust methods such as* $H_\infty$ *optimal control,* $\mu$*-synthesis and their extension to Linear Parametrically Varying (LPV) control are explored and demonstrated through several aeronautical design cases.*

## 1.1 Active Flight Control Technology

Space and flight have been technology drivers and have led to many innovations still influencing many non aerospace related industries. It is noteworthy that in the sixties, the technology acquired from the Apollo program was transferred on an almost one to one fashion into the F-8 Fly-by-Wire (FBW) demonstrator to explore the possibilities given by digital control. The FBW concept allows flight without direct mechanical linkage between pilot inceptors and the control surfaces.

The FBW technology provided the infrastructure for embedding flight control systems for the artificial augmentation of the aircraft dynamics. A feedback system is used to primarily stabilize the airframe while ensuring well-behaved dynamic aircraft responses to control inputs and disturbance. The description and flight test results of the F-8 FBW control system are documented in [Ano75]. More than FBW alone, AFC technology allows to accommodate for disturbance rejection, control energy minimization and accurate reference tracking in almost any dynamical aircraft subsystem.

In the field of Active Control Technology ACT, we recognize a blend of different disciplines i.e. Flight Mechanics, Flight Propulsion, Aerodynamics and Aircraft Structures. It is common that separate departments or groups represent the mentioned fields. This is often the case in the universities, the research organizations as well as in the aircraft companies. Although, they do have their specific philosophies, backgrounds and approaches to problem solutions, there is a need for an integrated design approach. It is evident that all parties involved in the design of an integral aircraft system should use common and compatible modeling, analysis and synthesis tools. System and Control theory aims at reaching these tools to the aeronautical community allowing to bridge the interdisciplinary gap documented in the overview paper given by Pelegrin [Pel85].

### 1.1.1 Evolution of Control Augmentation Systems

The primary function of a Control Augmentation System (CAS) is to stabilize the airframe dynamics. Tracking characteristics are obtained using appropriate feedback signals. For good longitudinal Handling Qualities (HQ) as specified in MIL-STD-1797, MIL-F-8785 C and Gibson [MIL87, MIL80, Gib82], the short period or pitching mode of the airframe, controlled by the elevator, is most important. There are many strategies for shaping the aircraft's short period response but three main can be distinguished, i.e. angle-of-attack, pitch rate, and normal acceleration command. The most widely adopted strategy is the pitch rate command system (PRCS) as described in detail in Gibson [Gib99]. Proportional and integral feedback of the pitch rate signal provide the pilot a PRCS when the stick is displaced and an attitude hold function when the stick is kept centered.

In the absence of direct lift devices the angle-of-attack is an uncontrolled variable. An angle-of-attack limiting function is necessary in the CAS system to prevent the pilot from stalling the aircraft. Physically, this is realized by introducing in the control system an angle-of-attack limiter.

Second, command shaping is provided by means of a feedforward control system that cancels and shifts known dynamics. Primarily the short period numerator zeros are moved while the effects of flexible dynamics are attenuated by notch filters. The two degrees of freedom control structure comprising the feedback and feedforward control functions is called in the flight control community the

Superaugmented CAS (SCAS) system. Feedforward is used to control all what is known as where feedback is used to account for the unknown. A good overview of the perspectives on superaugmented control are given by McRuer [RJM85]. Often a rate limiting element is introduced at the actuator in order to prevent too fast surface deflections.

For military aircraft it has been shown that static stability by design can be a limiting factor for high performance maneuvering tasks. The successes in the eighties of many NATO ACT research programs such as the Jaguar, F-104 Control Configured Vehicle (CCV), EAP later called EFA, Advanced Fighter Technology Integration AFTI/F-16, F-18, F-111 and many more are well documented in the AGARD proceedings [AGA85]. By means of Active Control Technology it has been shown that maneuvering capabilities can be enhanced if the airframe is allowed to have zero or marginal stability while ensuring stability with the aid of a full authority active control system.

Moreover, if direct force control and thrust vectoring are applied a variety of new maneuvers can be generated while the designers have more freedom for shaping the desired Handling Qualities. A detailed review on the aerodynamics of controls is given in the overview paper by Young [You85] and the references therein. As reported by Kaul, Evans, Korte, Daley and Toles in [HJK85, Eva85, Kor85, Dal85, Tol85], ACT allows, by virtue of feedback, to incorporate into one single airframe optimized and multirole maneuvering capabilities. The resulting airframe, depending on the mission to be flown has now artificial adaptive dynamic and aerodynamic characteristics.

### 1.1.2 Structural Mode Control and CAS's

As performance requirements and weight reduction constraints become more stringent, airframes have the tendency to become more flexible. Structural effects and distortions start to play an important role and interact with the rigid body dynamics. These effects will have noticeable design consequences, certainly in the latest generation of large civil aircraft such as the Airbus A380.

The fundamentals of aeroelasticity are described in textbooks by Displinghoff, Dowell and Foersching [BAH55, DI88, Foe74]. The aeroelastic influences can be categorized as a combination of aerodynamic, elastic, inertial and control forces. The aeroservoelastic effects on the stability and control of aircraft can be substantial. Moran [Mor85] reports that due to structural effects a total redesign of the originally conceived CAS system was necessary. For a historical overview and the implications of aeroelasticity in flight testing we refer to the papers given by Kehoe [Keh95].

In response to structural problems the B-52 LAMS (Aircraft Load Alleviation and Mode Stabilization) program and the B-52 CCV [RHF75] provided successful concepts such as the Identical Location for Accelerometer and Force (ILAF)

and the phase stabilization concept. This resulted in four mode control systems, such as the Ride Control system (RC), the Flutter mode Control system (FC), the Maneuver Load Control system (MLC) and the Augmented Stability system (AS).
Further, the original LAMS system was adapted to provide in an Fatigue Reduction system (FR). The flutter control system was flight-tested and allowed the aircraft to fly 10 knots above flutter speed proving in this way flight envelope extension and demonstrating the benefits of active control technology.

Although, a large number of benefits of active control technology was convincingly demonstrated, inherent limitations were reported. For instance, the amount of damping achievable by the use of the SISO phase stabilization concept is still poor. Also, the achieved SISO SMC bandwidth is still too low. Further pushing the SISO SMC bandwidth to higher levels leads to unwanted structural excitations or even to aeroservoelastic instabilities. These limitations are mainly due to inaccurate aeroelastic modeling and to unwanted interactions between the SMC and the CAS.

Even with the adoption of an integrated design approach to cure the interaction problems Schwanz [SB84] reports on the importance of accurate aeroelastic modeling for the design of integrated structural and CAS control systems. Schwanz shows what modeling assumptions on flexible structures give raise to substantial system uncertainties and model mismatch. It is shown how these modeling assumptions lead to series of interferences between the aeroelastics and the control system. Schwanz summmarizes for ten fighter aircraft in [SB84] various airframe vibration problems raging from flutter to limit cycles. It is shown where and how in aircraft structures these problems physically originate and if possible how these problems were solved. The problems were mostly cured by relaxing the control performance. This is achieved by reducing the control gain and control bandwidth to increase stability and the introduction of notch filters to cancel unwanted resonnances.

### 1.1.3 Autopilots and CAS's

Whereas many military aircraft rely on active stabilization, civil aircraft are mainly designed to be statically stable even though the benefits of unstable airframes are well known and understood.
Several civil aircraft relaxed stability studies on Airbus and Lockheed L-1011 as reported by Graeber and Grising [Gra85, JR85] demonstrate that reduced trim drag results in an increase of the flight range at a substantial reduction of operational costs. Relaxed stability necessicates an augmentation system to give good Handling Qualities, nevertheless, the concept has not been adopted in commercial aircraft.

In autopilot systems the CAS system is primarily used for inner loop pitch rate damping of the short period motion and accommodation of a basic attitude hold mode function for level flight at constant speed. The second longitudinal aircraft mode, much slower than the short period dynamics characterizes the point mass force equilibrium dynamics of the aircraft. In equilibrium flight the potential and kinetic energy levels are constant. These can be controlled using altitude and speed related commands. Augmentation of the primary CAS provides autopilot modes such as the flightpath or altitude hold mode. To keep the energy balance in an altitude select mode a separate autothrottle function is added. This mode compensates for speed changes.

Many problems were reported on the lack of overall performance of separately and successively designed autopilot/autothrottle modes and systems [Lam83c, Lam83d, Lam83a, Lam83b]. Similar problems on the classically designed flight propulsion system for the YC-14 were reported by Lee in [LKS78]. Moorhouse [MS85] quotes Lee with the following:

> *A completely satisfactory version of the speed hold system was never achieved. The interrelationship of pilot commands, speed hold system responses, sink rate and ground effect prior to touchdown were not completely understood.*

Lambregts [FL99] summarizes that the SISO loop at the time designed integrated autothrottle/autopilot systems mostly have unacceptably high control activity to wind gusts, lack of coordination and high levels of interaction between the flightpath and the speed command loop. It has been reported that for all B-747, the autothrottle speed command systems tend to cause an unstable flightpath response while the elevator commands at constant thrust induce unacceptable speed deviations. Furthermore, these systems showed high vulnerability to turbulence and windshear leading to unacceptable ride quality levels and poor tracking performance. The problems reported have been alleviated after costly system redevelopment in the later B-747-200/300 series by eliminating the flithpath loop to operate on the remaining attitude hold mode. Furthermore, by keeping the control bandwidth low in the autothrottle loop, safe stability margin levels were ensured at the cost of reduced performance.

To overcome the problem of coupled dynamics a series of integrated design philosophies have been proposed. One of them is the integrated autopilot-autothrottle system called the Total Energy Control System (TECS) [Lam83d] for the simultaneous control of the longitudinal motion with elevator and throttle. For the control of the lateral dynamics the Total Heading Control System (THCS) was proposed. These modes greatly simplified the traditional autopilot structure while still providing in all necessary functions for guidance. Gain scheduling as well as implementation of envelope protection function were greatly simplified while the amount of mode switching logic was drastically reduced. The resulting system had good coordinated and decoupled control, much higher tracking performance levels while being less sensitive to turbulence. The overall system provided also in manual control functions referred to as the Velocity Control Wheel Steering CWS control concept. Flight tests demonstrated in the automatic as well as in the manual mode good flying and handling qualities levels. Even though, these systems are technically superior to the single loop designs they have never been commercially adopted.

### 1.1.4 Towards more Complexity

Active control technology allowed the introduction of a whole new array of motivators. There are surfaces that account for load alleviation, ride control and envelope protection. Foreplane slats are used in the Eurofighter to provide adaptive aerodynamic lift to drag ratio optimized flight. In the X-31, F16-VISTA and HARV programs thrust vectored nozzles are used to extend the traditional flight envelope by providing enhanced low speed/high angle-of-attack maneuverability capabilities. Thrust vectoring was also used to provide Vertical Short Take-Off and Landing (VSTOL) capabilities for the Harrier.
Tailless flight is made possible by using thrust vectoring for lateral control. Current investigations concentrate on low signature tailless aircraft concepts on the Tailless Advanced Fighter Aircraft (TAFA). One of the main technological issues focusses on the demonstration of Reconfigurable Control for Tailless Fighter Aircaft (RESTORE) [BW98] using modern control concepts.

The F-15 Active program resulted in an demonstrator aircraft with numerous motivators raising new opportunities and challenges for multivariable control. New challenges such as the control allocation problem for which theoretical results are sparse are encountered. The main question is, how to manage and optimally allocate multiple redundant controls without saturating the motivators. This led to new control concepts such as the generalized moments control with reallocation schemes such as daisy chaining and pseudo-inverted control [DB96, Buf97].

In recent years several Fault Tolerant Control research projects have been carried out to extend system capabilities in case of faults or damage. The Propulsion Controlled Aircraft (PCA) concept is one example of Fault Tolerant Control applications. It emerged as a control solution to the Sioux City accident in which a DC-10 lost all hydraulic power for aerodynamic control. The crew managed to crash-land the aircraft using engine thrust only. Similarly the incident in 1977 where a Lockheed L-1011 experienced an elevator failure, pitch control was regained by using differential thrust. NASA Dryden Flight Research Center developed based on conventional design methods high performance propulsion control schemes to control the aircraft's flight path and heading. Burcham etal. [BBMF97] report that the propulsion control concept been successfully flight tested on the F-15, the MD-11 and as well as on the B-747 aircraft. The PCA system is only dedicated to cope with an anticipated system fault namely, the loss of hydraulic power for the elevator system. It is not suited to account for more general combinations of unexpected separate failures.

A more flexible and general FTC strategy which does not anticipate on a fault scenario is self-repairing control. Self-repairing control allows to reconfigure the control system for the degraded situation provided that enough control power is available. This strategy necessitates on-board computational power and algorithms to perform online system identification, control system redesign and control allocation. These issues have been investigated within the X-36 and RESTORE programs. Full advantage is taken of current modern robust and nonlinear multivariable tools. Recent flight-tests on the X-36 [Wal97] research vehicle have demonstrated the technological feasibility of the proposed concepts.

## 1.2 The Design and Certification Process

The flight control system design process can be seen as series of successive design activities of increasing complexity that in the end must lead to a flight-tested design. As the design evolves considerations made from flight mechanics, handling qualities, aerodynamics, atmospheric disturbances, structures, aircraft loads and human factors must all be integrated.

The efficiency of the design process depends on the modeling, the analysis and synthesis tools used and their computer implementation. These design tools should be *general* enough to cope in an uniform fashion with the multidisciplinary character of active flight control technology.

The subsequent steps in the control design process are given next.

### 1.2.1 The Flight Control Design Process

Every particular flight mission imposes a set of specific *requirements* on the system.
In the early stage of the system development the requirements are often of qualitative nature. For manually controlled military aircraft the requirement would impose features such as roll about the velocity vector for high angle-of-attack combat missions, or tight nose pointing for refueling mission. For autopilot systems one would require tight regulation of the reference altitude, flight path or vertical speed in the autopilot hold modes or tracking of these quantities in the autopilot select modes.
In this way the aircraft mission determines the choice for a control system architecture. This choice consists of the feedback arrangement of sensor arrays and actuator devices all properly calibrated and sized, connected to a control system in order to deliver desired aircraft dynamics.

Once the aircraft mission and the overall system architecture are fixed the actual control system design process starts. The process steps covered in this thesis are the following:

- **The modeling phase:** consists of a formulation of the design specifications. This comprises a mathematical development of both, a design model and the performance requirements the system must achieve.

- **The control-law synthesis phase:** on the base of the previously developed design model the next step is to synthesize a controller fulfilling the proposed design objectives.

- **The control-law analysis phase:** provides the designer the information whether the requirements are met or not. The trade-off between the various competing objectives must be revealed in order to refine the design when necessary.

In most cases a satisfactory control solution is only obtained after many iterations over these three design phases.

The most widely used design methods are only suited for LTI systems. An immediate consequence of this choice is that the design loop must be performed for each control mode over a set of fixed operating conditions in order to obtain adequate system behavior over the entire flight envelope.

Locally, which means at each fixed flight condition, the design has to meet the desired levels of single-loop stability margins and performance requirements. Linear time analysis is performed to confirm satisfactory performance robustness. Limiting functions such as position and rate limiters are often added afterwards to keep important signals within their physical prescribed boundaries. It is not uncommon that a designer has to conduct up to fifty local single-point designs to cover the flight envelope which is an artefact of using LTI systems and methods.

In the end, the individual control laws are scheduled as a function of key parameters (Mach number, aircraft mass, dynamic pressure, aerodynamic configuration) such that the resulting control law captures globally over the flight envelope the achieved LTI properties of each linear point design.

The scheduled control system is mostly validated by means of nonlinear simulations with the hope that the single point LTI design properties are preserved. In most cases the initial schedules and control laws need to be refined before satisfactory behavior is achieved. The performance robustness of the resulting design are commonly demonstrated by means of Monte-Carlo simulations.

These design steps are common for every seperate control system. In a similar fashion the structural group would design the SMC, the flight control group the CAS and autopilot, the propulsion group the engine control system and the flight mechanics group the navigation system. In the end all systems need to be integrated into one working concept.

### 1.2.2 The Evaluation Process

To validate the flying qualities fixed base and/or moving base simulator tests without and with hardware in the loop are performed in the subsequent control system evolution phase. Using pilot's opinion rated in Cooper Harper [CH69] scales as acceptability indicator new iterations between control design and piloted simulation are needed before the control system complies pilot's opinion. The origin of those many iterations lies in the mismatch between the mathematically formulated control requirements and the pilot opinion metric. Costly but necessary flight simulation hours, without and with hardware-in-the-loop considerably contribute to the overall system costs.

Finally, a flight test campaign needs to be performed to confirm the simulation results. It is more than normal that flight tests reveal system behavior unforeseen in the simulation phase. Often this is due to the substantial mismatch between the initially chosen design models and actual system. The design models need more than once to be adapted while the control laws need to be redesigned. The validation loop must entered again.

### 1.2.3 The General Problem

Benefits of active control technology have been widely demonstrated. Unfortunately, the sacrifices to master this technology are substantial. In this respect the Agard Working Group WG-17 has systematically analyzed the nature of the FCS design process. Some of Agard's WG-17 findings are quoted from [AGA91] to illustrate some of the problems and consequences encountered in the design of AFC systems.

> **The flying qualities of recently designed highly augmented aircraft have not always lived up to the hopes of their designers. The industry has seen some success, but has also encountered:**
>
> - *Loss of control during take-off in more than one instance*
> - *Loss of control in landing, in several instances ranging from identification of the problem in an in-flight simulator, to actual aircraft being damaged or even totally destroyed.*
> - *Difficulty in in-flight refueling, resulting even in airplane damage.*
> - *Expensively developed systems installed but remaining inactive, either because they failed to meet operational requirements or because they simply degraded the flying qualities they were supposed to enhance.*
> - *Total system redesign as almost a rule rather than an exception, increasing system cost manifold.*

- *Cancellation of an entire airplane project due to the expense and intractability of the augmentation system development.*
- *Removal of respected organizations from development teams because of stubborn resistance of the augmentation system to development progress.*
- *Loss of aircraft sales*

In response to these findings we may say that the origin of some of the above listed problems lies the complex nature of the design system for which not always adequate design tools are available. Most classical design tools are not *powerful* enough to cope entirely with the *complex* nature of current active control problems. The level of *complexity* in active control systems has several sources. Complexity increases as the system becomes more *multivariable*, more *nonlinear* and more *uncertain*. A further source of complexity is computational complexity which depends on the problem dimensionality. Some of the problems can be alleviated by taking complexity into account by design. This necessicates tools especially tailored the design task of complex AFC systems.

## 1.3 Classical versus Robust Setting

**The System**

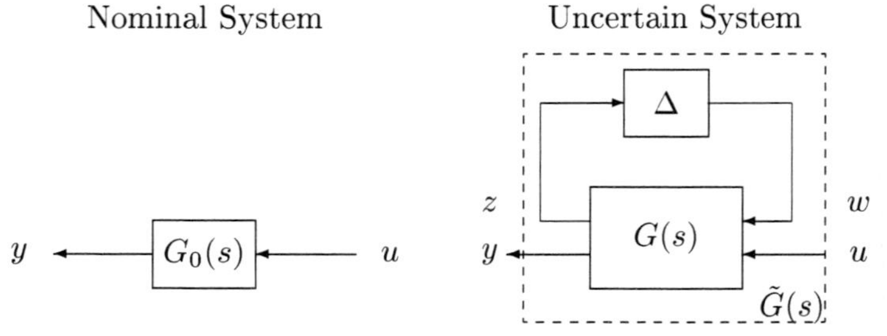


Figure 1.1: System Representation

Figure 1.1 depicts the two basic blocks representing the systems used in classical control and robust control.
In classical control we use the fixed LTI system $G_0(s)$ (left hand figure) which is equivalent to the uncertain set of systems $\tilde{G}(s)$ (right hand figure) whenever the operator $\Delta$ is set to zero. In robust control the operator $\Delta$ is deliberately introduced to represent the varying or the unkown elements the system. When the uncertainty $\Delta$ is set to zero the system $\tilde{G}(s)$ is said to be in the nominal

condition $G_0(s)$. The operator $\Delta$ can be a nonlinear, linear dynamical system, a time variyng or a constant parameter. Important is that the operator $\Delta$ is bounded in a norm $||.||_{-}$. For convenience the operator is normalized and may take any value within the interval $||\Delta|| \in [-1, 1]$.
To illustrate how the uncertain system $\tilde{G}(s)$ can be generated from the nominal system $G_0(s)$ and $\Delta$ we need the $2 \times 2$ block partitioned matrix $G(s)$.

$$\begin{bmatrix} z \\ y \end{bmatrix} = G(s) \begin{bmatrix} w \\ u \end{bmatrix} = \begin{bmatrix} G_{11} & G_{12} \\ G_{21} & G_{22} \end{bmatrix} \begin{bmatrix} w \\ u \end{bmatrix}$$

using the signal relation $w = \Delta\, z$ as shown in the figure 1.1 leads to

$$y = \tilde{G}\, u = \left[G_{22} + (G_{21}\Delta(I - G_{11}\Delta)^{-1}G_{12}\,\right] u$$

In the nominal case one sets $||\Delta|| = 0$. In that case the subblock $G_{22}$ corresponds to $G_0$. Assuming for simplicity that $G_{11} = 0$ and $G_{12} = G_{21} = I$ one obtains a simple set of systems given:

$$\tilde{G} = \{G_0 + \Delta\} \quad \forall\, ||\Delta|| \in [-1,\ 1]$$

There are infinitely many system conditions for $\tilde{G}$ when the norm of the operator $\Delta$ deviates from zero within the interval $[-1,\ 1]$. The connection of the system $G(s)$ with the operator $\Delta$ results in the set of uncertain systems denoted with $\tilde{G}$. The connection is called a Linear Fractional Transformation or LFT. LFTs form the basic system components in the robust control setting and generalize single LTI systems to sets of systems.

The operator $\Delta$ is based on engineering considerations and arises when we want to make explicit potential system variations. It might be that the system aerodynamics are poorly known due to the fact the aircraft operates in the transonic flight regime. The engineer has the possibility to model the variations expected in the aerodynamic coefficients.

In a similar fashion one can model at any system level other expected uncertainties. Alltogether these subsystems are connected yielding an interconnection of uncertain systems. The interconnection can be viewed as a new LFT of larger complexity with all local system uncertainties captured in a diagonally augmented operator $\Delta$. In this case we speak about structured uncertainties to make the distinction between unstructured uncertainties which can occur at any location in the system mathematically represented by an uncertainty operator $\Delta$ that is full.

### The System Interconnection

Next, consider the tracking problem as depicted in the left hand side of figure 1.2. The classical synthesis problem consist of designing a controller $K$ stabilizing the overal system. The performance objective is to achieve zero tracking error. The classical design rules consist of shaping the loop gain $G_0K$ over the different frequency ranges. At low frequencies high gain or integral action is added to get satisactory tracking behavior. In the mid-frequency range the cross-over frequency is increased to achieve faster system response. At high frequencies the system needs to have sufficient roll-off for robustness.

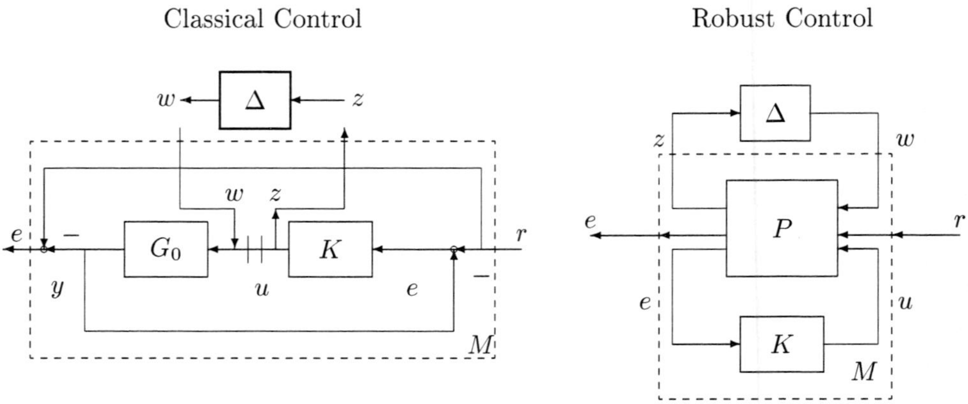


Figure 1.2: System Interconnection Representation

All these design rules are evaluated using graphical representations of the system dynamics within the Bode or Nyquist domain. In the classical setting checking the system margins is a sequential stability robustness problem to varying gain and/or phase. This test checks how much the compensator gain and/or phase are allowed to vary until instability occurs. This is equivalent to the introduction of a gain and/or phase uncertainty Δ at the actuator loop break point in the left side of figure 1.2 and then checking the margin.

The M and N circles represent the actual performance objectives of ideal system responses with according rise-time, overshoot and settling time characteristics. The nominal performance objective is met when the loop gain remains outside the closed-loop boundaries given by the M-N circles.

Next, by checking how much gain and phase variation remain in the loop gain until the boundaries in the M-N circle are violated is a robust performance analysis.

For multivariable systems the classical single loop margins are misleading robustness indicators. Classical gain and phase margins are not suited to predict the robustness properties of MIMO systems.

The robust control framework has been specially developed to overcome these shortcomings. The concept of breaking the loop for margin analysis has been generalized to the concept of physical uncertainty modeling Δ at any location in the loop.

The setting for robust control allows to deliberately introduce and generate systems with multiple simultaneous variations. In right hand side of figure 1.2 the equivalent of the classical tracking system architecture is redrawn. The elements consist of a system $P$, called the generalized plant, the controller $K$ and the operator $\Delta$, bounded in norm to reflect the system uncertainty. The system $P$ is called generalized plant because it includes all the design objectives and can be used for synthesis of a controller $K$ as well as for the analysis of the resulting closed-loop system $M$. The design objectives are captured by weighting functions similar to the boundaries given by the M-N circles.

The robust control synthesis problem is given the system $P$, find a controller $K$ stabilizing and optimizing in a certain norm the closed-loop transfer function from $(w, r)'$ to $(z, e)'$. Such an optimization problem is solved using either, Riccatti equations or Linear Matrix Inequalities depending on the chosen system norm and the performance objectives.

In the classical setting the controller is designed manually and is adjusted until visual inspection leads to satifactory performance. In the modern setting the design variables are weighting functions. For an achieved performance level a controller is automatically returned by the synthesis procedure. When performance is not satisfactory the design objectives captured by weighting functions included in $P$ need to be rescaled. Actually the design activity has shifted from controller synthesis towards specification synthesis.

Given $P$ and $K$ we can generate the general interconnection for analysis $M$ reflecting the controlled plant. The robustness analysis problem involves both $M$ and $\Delta$ while nominal analysis involves only $M$. Nominal stability is automatically achieved by the stabilizing controller. Nominal performance analysis involves in checking to what extent the prespecified performance objectives are met. Finally, the robustness analysis involves how well the nominal system properties are preserved under all allowable uncertainties given by $\Delta$. The study of worst case robustness properties involves special optimization tools which are reviewed next.

### 1.3.1 Modern Control and $\mu$-Synthesis

The ideas behind the $H_\infty$ theory were posed in 1981 by Zames [Zam81] and is motivated from the poor robustness properties of LQG regulators [Doy78]. Using the Singular Value loop shapes, Stein and Doyle [DS81] were able to extend the classical Bode loop shape design approach to the multivariable setting. In 1982 Doyle [Doy82, DWS82] generalized the Singular Value concept to the Structured Singular Value $\mu$ as a multivariable robustness indicator of multiply perturbed systems. These problems arise in mutliloop robustness analysis.

Starting from the $H_\infty$ robustness analysis given by the Small-Gain theorem re-

sults it has cost a decade to develop the $H_\infty$ controller synthesis results. The objective is to find a controller $K$, given $P$ such that the controlled system $M$ is robust against the worst case system unstructured perturbation $\Delta$. Practical and computationally efficient application of the $H_\infty$ control method was made possible by the formulation of state space solutions provided by Doyle and co-authors [DGKF88].
The extension to $H_\infty$ synthesis, primarly dedicated to systems with unstructured uncertainties, to the synthesis for diagonally perturbed systems is called $\mu$-synthesis. The idea is to optimize the upper bound of the structured singular value $\mu$. This $\mu$-upper bound is approximated by the $D$ scaled maximum singular value of $DMD^{-1}$.

The successes of $H_\infty$ and $\mu$-synthesis were also based on the availability of advanced numerical tools such as $\mu$-Tools [BDG+93b]. Engineering problems of significant size could be solved using $\mu$-synthesis making this a widely accepted design tool in the academic and research community.
Since then, a large amount of applied publications emerged demonstrating the benefits of the approach. Hyde [Hyd95] provided the first known flight tested $H_\infty$ control system for the VAC-Harrier.

### 1.3.2 Linear Matrix Inequality Approach

In the beginning of the nineties, attention has shifted towards a Linear Matrix Inequality approach to robust control. Boyd [BB91] motivated the interest because of the advances made in the field of convex optimization [NN93] from which control oriented software such as the LMI Control Toolbox [GNLC95], SDPSOL [BW95] and LMITOOL [GDN95] emerged.

On the other hand it was understood that the LMI approach provided the flexibility to simplify the formulation of many existing control problems as shown in [BGFB95].
An important result comes from the LMI formulation of the Small Gain theorem to which the $H_\infty$ analysis problem is related. This result is called the Bounded Real Lemma. Scherer [Sch90] directly relates the $H_\infty$-norm to the solution of Ricatti inequalities. Based on that formulation an LMI solution to the $H_\infty$ control problem was greatly simplified by Gahinet [GA94]. Using Linear Fractional Traformations (LFT), Doyle revisisted the problems related to the structured singular value $\mu$ from an LMI perspective [DPZ91]. In this view the paper by Packard *"A Collection of Robust Control Problems Leading to LMI's"* [PZPB91] provided the setting for the approach of a new class of problems such as the control of time varying systems and especially linear parametrically varying systems.
Using the LFT and $\mu$ machinery Packard and Balas [PZP+92] provided results for optimal constant input-output similarity scalings for the full-information and state-feedback control problems with time varying uncertainties. This problem was further extended by Packard by introducing a special LFT control structure

related to the gain scheduling problem. The synthesis solution of Gain Scheduling via LFT's was first provided by Parckard [Pac94]. Since then, this central result in LPV control was the object of continuous refinements [Hel95, GA95]. Based on an altnernative $H_\infty$ controller parametrization [GA94], Gahinet and Apkarian [GA95] provided a solution similar to Packard's gain scheduling problem. A more general controller parametrization was provided by Scherer [Sch95a, Sch95b] while the analysis results were generalized from the Bounded Real Lemma to the Differential Bounded Real Lemma. The Differential Bounded Real Lemma formed a powerful result allowing the treatment of the gain scheduling problem in a more general fashion and in a less conservative way than the LFT gain scheduling approach proposed by Packard. The solution to this problem relies on parameter dependent Lyapunov functions for which the LMIs can be solved using a gridding technique. These kinds of LPV problems were initially studied by Becker [Bec93] and extended by Wu [Wu95].

Scherer [Sch95a, Sch95b] generalizes and simplifies a large class previously obtained results The $H_\infty$ problem, the mixed $H_\infty/H_2$ problem, and extensions to the gain scheduling problem with constant and parameter dependent Lyapunov functions have been all reformulated using one single procedure.
Based on the analysis results given by Scherer in [Sch95b], Apkarian and Adams [AA96] refined the gain scheduling problem with parameter dependent Lyapunov functions together with parameter dependent multipliers to account for robustness issues in the gain scheduling problem.
A big step forward has been made in the analysis of interconnected systems since Rantzer and Megretski [MR97] introduced the notion of Integral Quadratic Constraints (IQC) allowing a genralized procedured for the search of multipliers. The existing analysis results such as the Kalman-Yakubovih-Popov Lemma, the Bounded Real Lemma, $\mu$ or even real-$\mu$ were easily derived and generalized using the IQC concept.
The combination of two powerful results, one for the analysis on LFT systems based on an IQC argument together with the generalized synthesis procedure delivered by Scherer [Sch95b] provided a general framework for design flexible enough to deal in the least conservative and straightforward way with the LFT gain scheduling problem for full-block scalings [Sch97].

## 1.4 Objectives and Approach

The evolution in robust control and the industrial needs for more efficient, reliable and better performing methods motivates the application of $\mu$-synthesis LPV methods to realistic flight control problems.

To demonstrate the features of the general framework several practical considerations related to the industrial needs have to be shown, these are:

- how flexible is the designer in the choice of the system architecture

- how flexible is the designer in the inclusion of specific aeronautical design specifications
- how good is the design solution
- how flexible the method is with respect to redesign in the face of evolving design requirements.

The controller design problem is characterized by a number of fundamental trade-offs between conflicting design specifications. The objective is to demonstrate how to incorporate the following objectives for the synthesis of flight control systems:

- **performance**: reflecting tracking behavior and disturbance rejection characteristics of certain signals;
- **robustness** : reflecting the stability bounds with respect to parameter variations;
- **ride quality**: reflecting the desire to obtain sufficient passenger and pilot comfort in the form of bounds on certain maximum allowable accelerations and minimum damping levels;
- **safety**: these reflect the envelope safeguards such as stall protection;
- **control activity**: as a measure of the power consumed by the controls giving also an indication of structural fatigue effects.

Further, we shall demonstrate how to use the several Structured Singular Value analysis tools to answer the question wether the designed system robustly achieves the requirements. This means that the achieved criteria for performance, ride quality, safety and control activity should be maintained for all possible parameter variations and system uncertainties.
As a final validation, the inherited performance robustness will be evaluated in the time domain by means of nonlinear Monte-Carlo simulations.
Demonstrating the design considerations leading to the fulfillment of the design requirements within the $\mu$ framework reformulates the first objective of this thesis.

The second objective in this thesis is to overcome the limitations inherent to Linear Time Invariant methods in general. We are interested in developing and applying formal extended design concepts for Linear Parametrically Varying systems. The motivation comes from the fact that in the aeronautical industry designing and implementing Gain Scheduled controllers is common practice. It is seen as a more involving design task than the linear design. Classical Gain Scheduling relies on engering skills. By using traditional interpolation schemes the resulting controller is not guaranteed to perform well within the defined operating range. With formal methods we aim a delivering control solutions with labeled performance and robustness guarantees.

The gain scheduling methods demonstrated in this thesis need a design approach similar to $\mu$-synthesis. These methods naturally extend the design features robust control techniques offer to the more general class of LPV systems. This motivates the second objective of this thesis and translates in showing how the design features acquired by $\mu$-synthesis can be extended to the gain scheduling problem.

# 1.5 Outline of the Thesis

To demonstrate the generality of the robust control framework several design procedures are illustrated on a variety of aeronautical control problems:

- Control Augmentation Systems meeting Handling Quality Requirements (Citation)
- Integrated Autopilot Systems for a Civil Aircraft (RCAM)
- Missile Autopilot System (Missile)

Problems with increasing complex specifications are solved. This is done to show how specific design considerations can be gradually included in the design. The following aeronautical control problems are solved:

- *Single-Loop LTI Robust Performance Handling Quality problem for the Cessna Citation using $\mu$-synthesis*
- *Multivariable LTI Robust Performance Autopilot problem for the RCAM using $\mu$-synthesis*
- *Single-Loop Gain Scheduled Robust Performance problem for a Missile using Formal Robust Gain Scheduled Control for Bounded Paramater Rates and for Unbounded Parameter Rates.*
- *Single-Loop Robust Performance LPV Handling Quality problem for the Cessna Citation with Formal Robust Gain Scheduled Control for Unbounded Parameter Rates*

In chapter 2 we review some background theory on the Structured Singular Value $\mu$ necessary to perform a $\mu$-synthesis design. The theory previously reviewed is applied in chapter 3 for several robustness and performance scenarios for the design of a Superaugmented Control Augmentation System for the Cessna Citation aircraft. Design considerations are illustrated necessary to solve a typical Handling Quality problem for a Pitch Rate Command System. Moreover, it is shown how to design, in one step a Superaugmented Control Augmentation System that meets well-defined handling and flying quality characteristics while keeping the control acitivity bounded to avoid actuator saturation in the face of uncertain actuator dynamics and uncertainties in aerodynamic stability and control derivatives.

$\mu$-Synthesis is aimed at achieving a desired performance level in the face of a prespecified uncertainty level in the system, this is called the robust performance problem. This problem involves two subproblems, the robust stability problem and the nominal performance problem which altogether constitute the three elements of $\mu$-analysis. The problem is solved whenever the overal system is stabilized and meets the handling quality requirements for all potential system uncertainties.

The synthesis part of the problem relies on the DK-iteration scheme, allowing to optimize the upper bound of structured singular value $\mu$ as a measure for the achievable robust performance level in the system.
To overcome the often reported difficulties in connection with the lack of guidance in the selection of weighting functions we solve the robust handling quality problem for several performance and uncertainty levels. Based on a variety of design situations we highlight how $\mu$ can be used to manipulate the trade-off between in the system available performance and robustness.

Next, in chapter 4 a $\mu$-synthesis autopilot design is performed on the RCAM problem. First we develop a large set of performance and robustness design requirements necessary for the autonomous approach to land of a civil aircraft. It is shown how to translate sets of multivariable design requirements to obtain TECS like command strategies for the autopilot system. It is demonstrated how to obtain at a reduced design effort using $\mu$-synthesis a TECS like control system with certain robustness guarantees. The robustness stability properties of the resulting control solution are validated by means of a $\mu$ analysis whereas the robusntness performance properties are revealed by means of nonlinear Monte-Carlo simulations.

Then chapter 5 discusses an approach to the control of Linear Parametrically Varying (LPV) systems that can take the rate of the scheduled parameter variations into account with robustness guarantees against parametric and dynamic uncertainties. The technique is illustrated on a missile control problem that has been extensively studied in the literature. The method is benchmarked and compared with other existing formal LPV techniques. The proposed method turns out to be the least conservative for this case study. The control solutions has a better performance measured in the $L_2$-gain. Using nonlinear simulations the inherited robustness properties are validated and confirmed.

In chapter 6 we use a standard Linear Parameter Dependent Linear Fractional Transformation (LPV-LFT) gain scheduling approach combined with an intermediate $\mu$-analysis step to take stability robustness requirements into account. More than only designing a robust gain scheduled controller it is shown how Level 1 Handling Quality requirements formulated in terms of a paramater dependent Gibson Control Anticipation Parameter (CAP) can be efficiently incorporated into the design. The properties of the resulting control system are analyzed over the flight envelope using the nominal performance (NP), robust

stability (RS) and robust performance (RP) indicators.
Finally, conclusions and recommendations for future research are given in chapter 7.

# Chapter 2

# Preliminaries: $H_\infty$ & $\mu$-Theory

*In this chapter we review the main results from robust control theory necessary to conduct a $\mu$-synthesis design. Using operators and operator norms we characterize the stability robustness property of uncertain interconnected systems. Then, we give tight guarantees for the robustness performance property for a system using the main results from $\mu$-analysis. Finally, we review how these analysis results are reformulated in a computational scheme, the so called $D-K$-iteration for the synthesis of robust performing control laws.*

## 2.1 Introduction

The research effort in the last decade devoted by the control community [DGKF88, Fra87, GD88, ZK88, KPR88, SLC89, Sch90, SMN90, GA93, SD91, LZD91] resulted in the robust control design framework that is general enough to handle a large class of realistic control problems relevant to practical engineering.

The robust control framework associates a general system representation and certain system "norms" to form the fundament for modern control synthesis and analysis. The general class of systems, called *Linear Fractional Transformations* has been studied in the early sixties by Redheffer [Red60]. It allows to model sets of uncertain systems in which a norm bounded perturbation operator affects the nominal system in a feedback way. The stability of such perturbed systems is maintained whenever the system outputs remains bounded under all perturbations.

The worst case amplification in the system is one such system norm and is given by the $H_\infty$ norm. The $H_\infty$ norm allows us to measure the degree of stability robustness in the system. The $H_\infty$ norm minimization problem, is to find a stabilizing controller minimizing the worst case system amplification. The

Structured Singular Value $\mu$ introduced by Doyle [Doy82, PD93] with reciprocal the multivariable stability margin [Saf82] refines the robustness analysis to systems affected by uncertainties that are structured. Further $\mu$-Synthesis introduced by Doyle during the 1984 ONR workshop [DCF⁺84], extends the $H_\infty$ optimization problem to systems with structured uncertainties. In this preliminary chapter a review is given of the theoretical results necessary for practical design.

## 2.2 Interconnected Systems

The performance robustness problem is addressed in a general framework for system design, which consists of a general problem description in terms of LFT systems, some key analysis results, a suitable measure of the magnitude for matrix transfer functions, the structured singular value $\mu$, and some controller synthesis results. An important remark is that all elements considered in this design framework have a practical software implementation in order to be useful for the engineering world. This issue is provided by the excellent and reliable software of [BDG⁺93a]. Furthermore, the practicing engineer needs good tutorials to keep track of the theoretical advances in the field, these are nowadays richly provided in references as [BDG⁺93a, PDB93, ZDG96, DFT92] and many others. The various elements of that framework are described next.

**General System Interconnection**

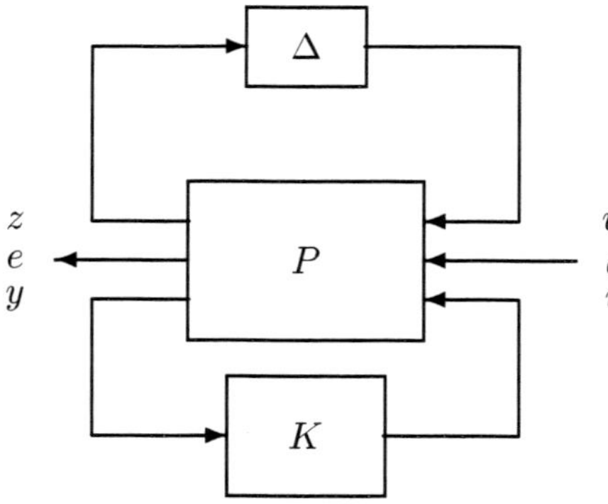


Figure 2.1: General System Interconnection

Using the LFT representation the classical multivariable control problem can be transformed into a more versatile form as shown in figure 2.1. This system representation together with $\mu$ allows to formulate tight bounds for the robust performance problem.

We recognize for any uncertain closed-loop system the three basic components:

1. **The General System $P$ (our problem data)**
2. **The Controller $K$ (possibly still to be designed)**
3. **The System Uncertainty $\Delta$ (belonging to a prespecified set).**

All uncertain elements have been pulled out of the system and placed in what is called a $\Delta$-block. The only assumption made on $\Delta$ is that is a stable and norm-bounded operator with $||\Delta||_\infty \leq 1$.

Three sets of input-output signals pairs are involved in the design of robust control-laws:

- **The Set of Control and Measurements Signals $(u, y)$ with $u = Ky$.**
- **The Set of External Commands, Disturbances and Performance Variables $(d, e)$.**
- **The Set of Perturbation Signals $(w, z)$ for which $w = \Delta z$.**

In terms of robust flight control-laws these signals can be interpreted in the following way:

- the control signals $u$ can be surface deflections and thrust controls
- the measurements signals $y$ are measured aircraft variables, given by the states, their combination from an on boeard sensor device
- the external commands and distubances $d$ can be pilot inputs or references from a flight management system. The disturbances can be either wind gusts or sensor noises.
- the performance variables $e$ are signals the designer wants to watch and to keep small in a certain norm. These can be tracking errors, surface deflections and surface rates. Another example is the angle of attack, the designer may penalize this signal to prevent aerodynamic stall. In principle any other measurement important for a well behaved system operation can be used.
- the perturbation input signals $w$ are introduced by the designer at system components in the loop where potential problems are foreseen. This can be at the actuator, the sensor or any potential system location subject to variations.
- the ouput perturbation signals $z$ are introduced by the designer to watch the effect of the signal injection $w$ at the actuator, sensors or any other system location.

The elements of the generalized plant $P$ consist of a $3 \times 3$ block-structured matrix according to the signals from $(w, d, u)'$ to $(z, e, y)'$:

$$\begin{bmatrix} z \\ e \\ y \end{bmatrix} = \begin{bmatrix} P_{11} & P_{12} & P_{13} \\ P_{21} & P_{22} & P_{23} \\ P_{31} & P_{32} & P_{33} \end{bmatrix} \begin{bmatrix} w \\ d \\ u \end{bmatrix} \tag{2.1}$$

together with the matrix $K$ that relates $y$ to $u$ as $u = Ky$ and the operator $\Delta$ mapping $z$ to $w$ according to $w = \Delta z$.

The design problem is to find a compensator $K$ internally stabilizing the general system $P$ while keeping the matrix transfer function between $d$ and $e$ appropriately small for the whole set of allowable perturbations $\Delta$. In the transformation process, from the classical setup into the general setup, any uncertainty arising at system component (actuator, plant sensors etc.) level becomes automatically structured at the level of the generalized system $P$. Furthermore, the so obtained generalized problem description as given in figure 2.1 is suitable for the synthesis problem as well as for the analysis problem.

Closing the lower channels of $P$ with $K$ involves a *Lower Linear Fractional Transformation* of $P$ by $K$:

$$\begin{aligned} \begin{bmatrix} z \\ e \end{bmatrix} &= F_l(P, K) \begin{bmatrix} w \\ d \end{bmatrix} \\ &= \left( \begin{bmatrix} P_{11} & P_{12} \\ P_{21} & P_{22} \end{bmatrix} + \begin{bmatrix} P_{13} \\ P_{23} \end{bmatrix} (I - P_{33}K)^{-1} \begin{bmatrix} P_{31} & P_{32} \end{bmatrix} \right) \begin{bmatrix} w \\ d \end{bmatrix} \\ &= M \begin{bmatrix} w \\ d \end{bmatrix} \end{aligned} \tag{2.2}$$

the index $l$ in $F_l(P, K) = M$ indicates that the *lower* channels of $P$ have been closed with $K$.

The underlying assumption made is that the nominal system $P_{33}$ is strictly proper, the compensator $K$ is proper such that $K$ stabilizes $P$. This assumption ensures that $M$ is stable and that indeed the nominal closed-loop transfer function $(I - P_{33}K)^{-1}$ is stable.

The stable system $M$ has a $2 \times 2$ block structure and is used to analyze the stability and performance robustness properties of the closed-loop system.
The general expression for each block $M_{ij}$ in the terms of $P$ is given by:

$$M_{ij} = P_{ij} + P_{i3}[I - KP_{33}]^{-1}KP_{3j} \quad i,\ j = 1,\ 2 \tag{2.3}$$

We see that since $K$ stabilizes the plant $P$ indeed all blocks $M_{ij}$ are stable.

To analyze the robustness properties of the loop introduced by the operator $\Delta$ we close the *upper* channels of $M$ with the appropriately dimensioned matrix $\Delta$

relating $w$ to $z$. The upper LFT is given by:

$$\begin{aligned} e &= \left[M_{22} + M_{21}\Delta\left(I - M_{11}\Delta\right)^{-1} M_{12}\right] d \\ &= F_u(M, \Delta) d \end{aligned} \tag{2.4}$$

The conditions to ensure that $F_u(M, \Delta)$ behaves well are given next.

## 2.3 Basic Stability Results

Figure 2.2 reflects the prototype configuration for the stability robustness analysis problem. The problem is, given the stable system $M$, determine how large the perturbation $\Delta$ can become before the closed-loop system becomes unstable.

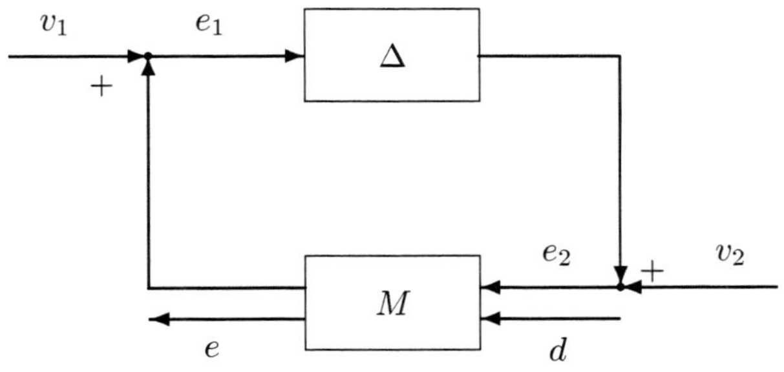


Figure 2.2: Generic Feedback Strucure

This structure allows to study depending on the situation imposed by the nature of the operators $M$ and $\Delta$ stability robustness problems. Stability of a linear time-invariant system is determined by location of all poles in the left half plane. Robust stability in the is determined by considering an interconnection of stable operators i.e. $M, \Delta \in \mathbb{R}\mathrm{H}_\infty$.

The operator $\Delta$ is chosen to be a bounded operator $||\Delta||_\infty < 1$. The operator $\Delta$ designates parts of system dynamics that are subject to potential changes. These are deliberately introduced into the system by the constraint $w = \Delta z$ to study their effect on the stability properties of the system.

To insure that the interconnection between $M$ and $\Delta$ is well behaved the signals $(e_1, e_2)$ need to be bounded for any bounded signals $(v_1, v_2)$.
Observe in the expression for $F_u(M, \Delta)$:

$$\begin{aligned} e &= \left[M_{22} + M_{21}\Delta\left(I - M_{11}\Delta\right)^{-1} M_{12}\right] d \\ &= F_u(M, \Delta) d \end{aligned}$$

the operator $\Delta$ only affects the $M_{11}$ block relating the internal signals, hence the stability properties of $M$ involve only the matrix $(I - M_{11}\Delta)^{-1}$ to be well behaved.

**Theorem 2.3.1** *[General Nyquist] Given the feedback interconnection structure of figure 2.2 for stable transfer function operators $M, \Delta : L_2 \to L_2$ with $M, \Delta \in I\!R\mathrm{H}_\infty$, if for all $s \in jI\!R$ the Nyquist locus does not become singular, i.e., $det(I - M_{11}\Delta)(s) \neq 0$*

1. *the closed-loop system is well-posed and internally stable and a unique $(e_1, e_2) \in L_2$ is associated with each $(v_1, v_2) \in L_2$*
2. *Furthermore, if the spectral radius $\rho(M_{11}\Delta) < 1$, then the closed-loop system is guaranteed to be stable for all $s$.*

The determinant condition given by the Nyquist theorem 2.3.1 provides a powerful mean to formulate the stability margin of the system $M$ with respect to norm bounded perturbations $\Delta$.

The second condition in the Nyquist theorem 2.3.1 is associated with the first condition to guarantee a well-posed and stable system.
The basic result for stability robustness analysis of interconnected systems as shown in figure 2.2 is the *Small Gain Theorem* [Zam66, Vid78].
The Small Gain Theorem states that a closed-loop feedback system of stable operators is internally stable if the loop gain of those operators is stable and bounded by unity.

**Theorem 2.3.2** *[Small Gain Theorem] Given the feedback interconnection structure of Figure 2.2 for stable transfer function operators $M, \Delta : L_2 \to L_2$ with $M, \Delta \in I\!R\mathrm{H}_\infty$, if the $H_\infty$-norm of the loop gain is bounded by unity such that $||M_{11}\Delta||_\infty < 1$, then*

1. *the closed-loop system is well-posed and internally stable*
2. *a unique $(e_1, e_2) \in L_2$ is associated with each $(v_1, v_2) \in L_2$*

The Small Gain theorem allows us to analyze robust stability of a plant model with respect to a set of perturbations. These perturbations describe plant uncertainty due to model errors and unmodeled dynamics. Often the modeling error is not exactly known but bounded by a norm $\gamma > 0$. Let $\Delta$ be the set of norm bounded operators describing the perturbations affecting the generalized plant for anlaysis:

$$\Delta = \{\Delta : ||\Delta||_\infty \leq \gamma\}$$

The Small Gain theorem involves the $H_\infty$-norm of the loop gain $||M_{11}\Delta||_\infty$. The condition for robust stability can be obtained using:

$$||M_{11}\Delta||_\infty \leq ||M_{11}||_\infty ||\Delta||_\infty$$

The plant $M$ and hence $M_{11}$ is robustly stable with respect to the set of uncertainty perturbations $\Delta$ which enter the system as in figure 2.2 with $||\Delta||_\infty \leq \gamma$ for all $\Delta \in \mathbf{\Delta}$ if

$$||M_{11}||_\infty < \frac{1}{\gamma}$$

The above condition is a sufficient, but not a necessary condition for robust stability. The Structured Singular Value $\mu$, is introduced in the next chapter as a less conservative measure of robust stability which is sufficient and necessary.

## 2.4 $\mu$-Analysis Results (Doyle 1982)

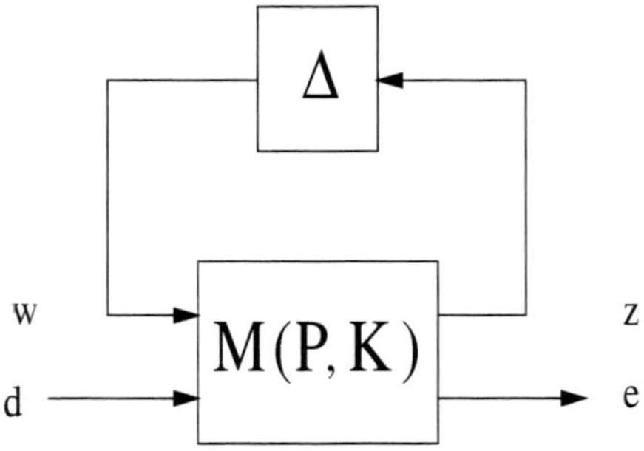


Figure 2.3: Closed-Loop System $M$

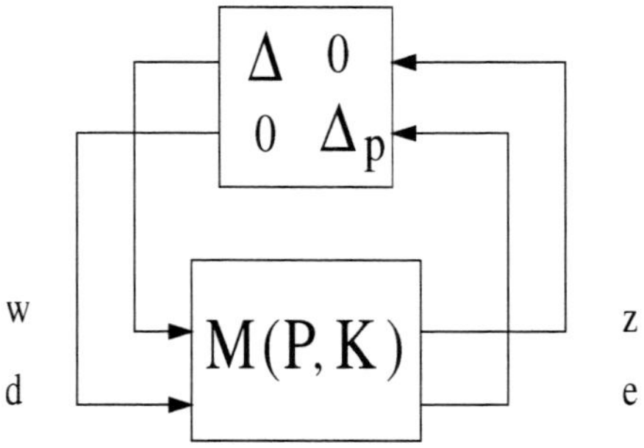


Figure 2.4: Stability configuration for Robust Performance

From the general system representation as shown in figure 2.1 a nonconservative necessary and sufficient condition for robust performance can be derived. Given a controller $K$ nominally stabilizing the feedback loop $M = F_l(P, K) \in \mathbb{R}\mathrm{H}_\infty$ shown in figure 2.1 and the block-structured operator $\Delta \in \mathbb{R}\mathrm{H}_\infty$ with $||\Delta||_\infty \leq 1$ then the perturbed closed-loop system shown in figure 2.3 is given by is $F_u(M, \Delta)$. Using the stuctured singular value tight bounds can be established on how large the matrix $M$ can become to guarantee the stability and performance robustness properties in the face of block structured perturbations $\Delta$.

The stuctured singular value is a matrix function denoted $\mu(\cdot)$ on complex matrices $M \in \mathbb{C}^{(n_w+n_e)\times(n_z+n_d)}$. with an underlying structure $\mathbf{\Delta}$ depending on the uncertainty and the performance objectives.

The general uncertainty structure naturally arising from the LFT modeling process consists of $S$ Repeated Scalar blocks and $F$ Full blocks.

The dimensions of the repeated scalar blocks are gathered in the vector $r = (r_1, \cdots, r_i, \cdots, r_S)$ where the *i-th* scalar repeated block has the dimension $r_i \times r_i$. In the same fashion the dimensions of the full blocks are contained in the vector $m = (m_1, \cdots, m_j, \cdots, r_F)$ where the *j-th* full block has the dimension $m_j \times m_j$.

In this way the uncertainty $\Delta \subset \mathbb{C}^{(n_z+n_e)\times(n_w+n_d)}$ is defined as:

$$\Delta = \{ \text{ diag } \{\delta_1 I_{r1}, \cdots, \delta_{rS} I_{rS}; \Delta_1, \cdots, \Delta_F\} \ \delta_i \in \mathbb{C}, \ \Delta_j \in \mathbb{C}^{m_j \times m_i} \} \tag{2.5}$$

The uncertainty $\Delta$ is confined to be in a bounded subset $\mathbf{B}\Delta$ of $\Delta$ defined as:

$$\mathbf{B}\Delta = \{\Delta \in \Delta \text{ with } \overline{\sigma}(\Delta) \leq 1\} \tag{2.6}$$

The observation to view the performance robustness problem as a stability test brought the function $\mu(.)$ defined in equation 2.7. The function $\mu$ for complex matrices $M$ has been defined by Doyle [Doy82] in the following way:

**Definition 2.4.1** *The Structured Singular Value $\mu(M)$ is defined as:*

$$\mu(M) \triangleq \left[ \min \left\{ \epsilon \left| \begin{array}{l} \det(I - \epsilon\Delta M) = 0 \\ \textit{for some } \Delta = \mathbf{B}\Delta \end{array} \right. \right\} \right]^{-1} . \tag{2.7}$$

*The structured singular value $\mu$ is the reciprocal of the smallest value $\epsilon$ which makes the matrix $I - \epsilon\Delta M$ singular for some $\Delta$ in a block-diagonal set. When no $\epsilon$ exists, $\mu$ is taken to be zero.*

To motivation to call $\mu$ the *structured singular value* come from the fact the definition of $\mu$ reduces to the the maximal singular value in the absence of structure, that is when $\Delta$ is a full block.
The general condition for nominal performance is satisfied when bounded generalized disturbances $d$ are mapped into the set of bounded generalized error signals $e$. This is equivalent to keep the generalized weighted performance matrix $M_{22}(s)$ in an $H_\infty$ norm sense small. Nominal performance is achieved if and only if $\bar{\sigma}(M_{22}(jw)) < 1$ over the whole frequency range.

Robust stability involves the matrix $M_{11}$ and the uncertainty $\Delta$. Stability of the perturbed system $M_{11}$ is maintained as long as the matrix $(I - \Delta(s)M_{11}(s))$ does not become singular on the $j\omega$-axis for all $\Delta$'s. The Small Gain Theorem states that the robust stability is maintained as long as $\bar{\sigma}(M_{11}(jw)) < 1$ for all frequencies. The exact condition for robust stability with respect to block-diagonal uncertainty structures is satisfied if and only if

$$\mu_\Delta(M_{11}(j\omega)) < 1 \quad \forall \omega \tag{2.8}$$

for a given block-structured uncertainty from the set $\Delta \in \mathbf{B}\Delta$.

The robust performance condition as established in Stein and Doyle [SD91] is achieved if and only if the perturbed performance transfer function matrix $F_u(M, \Delta)$, remains in an $H_\infty$ norm sense bounded by unity, that is, if and only if:

$$\bar{\sigma}\left(M_{22} + M_{21}(I - \Delta M_{11})^{-1}\Delta M_{12}\right) < 1 \ \ \forall \omega, \ ||\Delta||_\infty \leq 1 \tag{2.9}$$

This norm condition can also be established by imposing the system $M$ in figure 2.4 to remain bounded in the face of a second norm-bounded perturbation $\Delta_p(s)$ from $e$ to $d$. In this way robust performance is equivalent to robust stability in the face of a block-diagonal structured perturbation, $diag(\Delta,\, \Delta_p)$, connected around the system $M$ as shown in figure 2.4. The system $M$ is robustly stable **if and only if** the function $\det(I - diag(\Delta, \Delta_p)M(j\omega))$ remains non-zero for all $\omega$. The determinant remains non-zero **if and only if** $\mu(M) < 1$ on the imaginary axis.

Under the condition the system $M = F_l(P, K)$ is nominally stabilized by $K$ the main analysis results are given:

**Theorem 2.4.2** *General Analysis Theorem* **Doyle**

1. *Nominal performance is satisfied* **if and only if**

$$\bar{\sigma}\left(M_{22}(j\omega)\right) < 1 \;\; \forall\, \omega \tag{2.10}$$

2. *Stability is robust* **if and only if**

$$\mu(M_{11})(j\omega) < 1 \;\; \forall\, \omega \tag{2.11}$$

3. *Performance is robust* **if and only if**

$$\mu(M(j\omega)) < 1 \;\; \forall\, \omega \tag{2.12}$$

## 2.5 Some Properties of $\mu$

In general exact computation of the structured singular value is not possible. Therefore, we work with approximations via the upper and lower bounds of $\mu$. For a complete overview on the complex structured singular value $\mu$ and the involved numerics we refer to [PD93, BDG$^+$93a]. The calculation of the structured singular value ($\mu$) relies always on the particular choice of the uncertainty structure $\Delta$ and generalizes matrix measures as the singular value and the spectral radius. For general uncertainty structures, the function $\mu$ is not a norm, since it does not satisfy the triangle inequality. Recall the complex matrix $M \in \mathbb{C}^{n \times n}$ with uncertainty $\Delta \in \mathbb{C}^{n \times n}$:

$$\Delta = \left\{ \text{ diag } \{\delta_1 I_{r1}, \cdots, \delta_{rS} I_{rS}; \Delta_1, \cdots, \Delta_F\} \; \delta_i \in \mathbb{C},\; \Delta_j \in \mathbb{C}^{\,m_j \times m_i} \right\} \tag{2.13}$$

then,$\forall\, \alpha \in \mathbb{C}\; \mu\,(\alpha M) = |\alpha|\, \mu\,(M)$. In some sense $\mu$ is related to how large the matrix $M$ is but it is not a norm. In special cases when $\Delta$ is either a repeated or full block uncertainty, the structured singular value boils down to common linear algebra measures such as the spectral radius $\rho$ and the maximum singular value $\bar{\sigma}$.

- Repeated Uncertainty $\Delta = \{\delta I \text{ with } \delta \in \mathbb{C}\}$ with $\| \Delta \| \leq 1$

$$\left\{ \begin{array}{l} S = 1 \\ F = 0 \end{array} \right. \text{ then } \mu_{\Delta}(M) = \rho(M)$$

- Full Block $\Delta \in \mathbb{C}^{n \times n}$

$$\left\{ \begin{array}{l} S = 0 \\ F = 1 \end{array} \right. \text{ then } \mu_{\Delta}(M) = \overline{\sigma}(M)$$

For a general $\Delta$-structure as in the definition of $\mu$ we get the following bounds.

$$\rho(M) \leq \mu_{\Delta}(M) \leq \overline{\sigma}(M) \tag{2.14}$$

Based on this generalization computable bounds can be given and refined. Define the following subset of $\mathbb{C}^{n \times n}$.

$$\mathbf{Q} = \{ Q \in \boldsymbol{\Delta} \colon QQ^* = I_n \} \tag{2.15}$$

$$\mathbf{Q} = \left\{ \begin{array}{c} \text{diag } [q_1 I, \cdots, q_S I; Q_1, \cdots, Q_F] \\ q_i \in \mathbb{C}, |q_i| = 1 \\ Q_j \in \mathbb{C}^{r_i \times r_i} \text{ with } Q_j^* Q_j = I. \end{array} \right\} \tag{2.16}$$

We see that the elements $Q$ of $\mathbf{Q}$ are unitary.
Define further the set of scaling matrices as:

$$\mathbf{D} = \left\{ \begin{array}{c} \text{diag } [D_1 I, \cdots, D_S I; d_1 I, \cdots, d_F I] \\ \text{with } D_i = D_i^* > 0 \\ \text{and } d_j > 0 \end{array} \right\} \tag{2.17}$$

Notice, that at places where the diagonal block of $\Delta$ is full the scaling matrix $D$ is a scalar, and vice versa. It can be seen directly that for each $D \in \mathbf{D}$ and $\Delta \in \boldsymbol{\Delta}$ the following important property holds, namely:

$$D\Delta = \Delta D$$

meaning that the scaling matrix $D$ communtes with the uncertainty $\Delta$.
Without proof the most important properties of $\mu$ for constant matrices $M$ are given next:

1. $\mu(\alpha M) = |\alpha| \mu(M)$
2. $\mu(I) = 1$
3. $\mu(AB) \leq \mu(A) \overline{\sigma}(B)$
4. $\mu(\Delta) = \overline{\sigma}(\Delta) \quad \forall \Delta \in \boldsymbol{\Delta}$

5. $\mu(M) = \mu(MQ^*) = \mu(QM) \quad \forall Q \in \mathbf{Q}$

6. $\mu(M) = \mu(DMD^{-1}) \quad \forall D \in \mathbf{D}$

7. $\rho(M) \leq \mu(M) \leq \overline{\sigma}(M)$

Using the transformations $D$ and $Q$ on $M$ the bounds on $\mu$ can be refined to:

$$\max_{Q \in \mathbf{Q}} \rho(QM) \leq \mu(M) \leq \inf_{D \in \mathbf{D}} \bar{\sigma}(DMD^{-1}) \tag{2.18}$$

For practical design the right inequality is important since the optimization problem of $\bar{\sigma}(DMD^{-1})$ is convex over the set scalings $D$. Furthermore, the optimization problem of $\bar{\sigma}(DMD^{-1})$ leads to an *upper bound* of the structured singular value $\mu(M)$.

The class of perturbations that has been considered here is related to complex matrices or scalars. For a complete treatment of the matematical properties of the complex structured singular value the interested reader is referred to [PD93].

A more specialized type of uncertainties is the set of the real parametric uncertainties arising in many flight control problems. This modeling type is especially well suited for the purpose of certification of control laws. In that case we would like to determine, given a flight control system with uncertain parameters, what is the smallest possible combination of parameter variations causing the system to become unstable. The structured singular value test for the set of real parametric perturbations is called the real $\mu$ test and is computationally more exhaustive than in the complex case. Nevertheless there exist for $\mu$ calculations with a mixed complex/real perturbation set, reasonably tight upper bounds which amount in finding optimal scaling matrices called the D-G scales, for more details the reader is referred to [ZDG96, BDG+93a].

## 2.6 $\mu$-Synthesis

The next step is the controller synthesis problem. The objective is to find a stabilizing controller $K$ achieving the desired performance requirements for the whole set of plants $\tilde{P}$.

$$\tilde{P} = \{F_u(P, \Delta_{pert}) \ : \ \Delta_{pert} \in \mathbf{\Delta_{pert}}, \ ||\Delta_{pert}||_\infty \leq 1\} \tag{2.19}$$

The definition of the $\mu$-**synthesis** objective:

**Definition 2.6.1** $\mu$**-synthesis :**
*Minimize over all stabilizing controllers $K(s)$ the worst case performance, i.e. the peak value of $\mu_\Delta(M) = \mu_\Delta(F_l(P, K))$.*

$$\min_{\substack{K(s) \\ stabilizing}} \quad ||[F_l(P, K)]||_{\mu_\Delta} < 1 \tag{2.20}$$

with the shorthand notation of the $\mu$-norm of the operator $G$ and similarly to the $\infty$-norm we have $\|G\|_{\mu_\Delta} = \max_\omega \mu_\Delta\,(G(j\omega))$ The structured singular value does not satisfy the definition of a norm. This notation is adopted only to reflect the fact that we want to measure the "size" of the worst case performance.
In order to perform calculations we replace $\mu_\Delta(\cdot)$ by its upper bound $\bar{\sigma}(D(\cdot)D^{-1})$. Define $\mathbf{D}_{pert}$, the scaling set for the perturbation structure $\Delta_{pert}$ . For $D_{pert} \in \mathbf{D}_{pert}$ and $\Delta_{pert} \in \mathbf{\Delta_{pert}}$ it follows from the definition of the invariance of $\mu$ under scaling, $D_{pert}\Delta_{pert} = \Delta_{pert}D_{pert}$, that the scaling set $\mathbf{D}$ for the augmented perturbation set is defined as:

$$\mathbf{D} = \left\{ \begin{bmatrix} D_{pert} & 0 \\ 0 & I \end{bmatrix} \; : \text{where } D_{pert} \in \mathbf{D}_{pert} \right\} \tag{2.21}$$

Note that the $D$-scale corresponding to the performance block $\Delta_p$-block is set to one. With respect to the scaling structure $D \in \mathbf{D}$ $\mu_\Delta$ can be obtained from the upper bound relation (applied to some constant matrix $M$):

$$\mu_\Delta(M) \leq \min_{D \in \mathbf{D}} \bar{\sigma}(DMD^{-1}) \tag{2.22}$$

When $\Delta_{pert}$ consists of $F$ full blocks, the set $\mathbf{D}$ looks like

$$\mathbf{D} = \{\text{diag}[d_1 I, \; \dots \; , \; d_F I, \; I] \quad d_j > 0\} \tag{2.23}$$

The elements of $D$ can have any phase without affecting the value of $\bar{\sigma}(DMD^{-1})$. The optimization along the frequency over $D$ can be replaced by an optimization over stable minimum-phase transfer functions $D(s)$. When considering real-rational, stable and minimum-phase scalings $D(s)$, the formulation of the actual optimization becomes:

$$\min_{\substack{K(s) \\ \text{stabilizing}}} \quad \min_{\substack{D(s) \in \mathbf{D} \\ \text{stable, min-phase}}} \quad \|D(j\omega)F_l(P,K)(j\omega)D^{-1}(j\omega)\|_\infty \tag{2.24}$$

In this way the synthesis problem that minimizes the worst case performance has been reformulated into a standard $H_\infty$ -synthesis problem. The simultaneous optimization over $D$ and $K$ is in general not convex. Therefore an indirect synthesis scheme is used to find a controller minimizing $\mu$. The procedure is called $D - K$-iteration since it iteratively optimizes over the stabilizing $K(s)$ while holding $D(s)$ fixed and then optimizes over stable minimum-phase $D(s)$ while holding $K(s)$ fixed. More details concerning the practical implementation of the synthesis problem can be found in [BDG$^+$93a, PDB93]. Most encountered practical engineering situations have revealed that the proposed scheme works well.

## 2.7 Conclusion

It has been shown that the $\mu$-design framework consists of 3 main elements. The first element is the LFT system representation. This is a general system

representation in which we can handle sets of uncertain systems in feedback configuration. Within this same representation we have seen the importance of the general open-loop interconnection structure $P$ and the general closed-loop system $M$.
The second framework element is the general $\mu$-analysis result on the closed-loop system $M$. This result allows to formulate the design solution in terms of robust stability, nominal performance and robust performance objectives.
Finally, the third element is the $\mu$ controller synthesis tool bridging design requirements into design solutions. The $\mu$-analysis results are used again to check wether the requirements are met or not. Accordingly, the requirements are tuned iteratively until the overall solution converges to an optimum level.

# Chapter 3

# $\mu$-Synthesis: Cessna Citation $\mu$-Design

*This chapter treats the application of $\mu$-synthesis for the design a Pitch Rate Command System. In section 3.1 the design problem is described and the performance and robustness objectives are formulated. Next, in the sections 3.2, 3.3 and 3.4 the design requirements are translated in a form suitable for analysis within the $\mu$-framework. Several design scenarios are performed to illustrate the effects of weighting function selection and uncertainty modelling. The synthesis results are summarized in section 3.5. In section 3.6 the obtained robustness properties are analyzed using $\mu$-analysis. Time simulations are performed in section 3.7.*

## 3.1 Problem Formulation: Pitch Rate Command System

The application of the $\mu$-theory as discussed in chapter 2 will be illustrated on the design of a Superaugmented Pitch Rate Command System.
The design plant is a linear model of the longitudinal short period dynamics of a Cessna Citation 500 in landing configuration. The model states are the pitch rate $q$ and the angle-of-attack $\alpha$. The state space representation of the model dynamics is given as:

$$\begin{aligned} \begin{bmatrix} \dot{q} \\ \dot{\alpha} \end{bmatrix} &= \begin{bmatrix} M_q & M_\alpha \\ 1 & Z_\alpha \end{bmatrix} \begin{bmatrix} q \\ \alpha \end{bmatrix} + \begin{bmatrix} M_{\delta_e} \\ Z_{\delta_e} \end{bmatrix} \delta_E \\ \begin{bmatrix} q \\ \alpha \end{bmatrix} &= \begin{bmatrix} 1 & 0 \\ 0 & 1 \end{bmatrix} \begin{bmatrix} q \\ \alpha \end{bmatrix} \end{aligned} \tag{3.1}$$

The input is the elevator deflection $\delta_E$. The output equation consists of the pitch rate $q$ and and angle-of-attack $\alpha$ measurements.

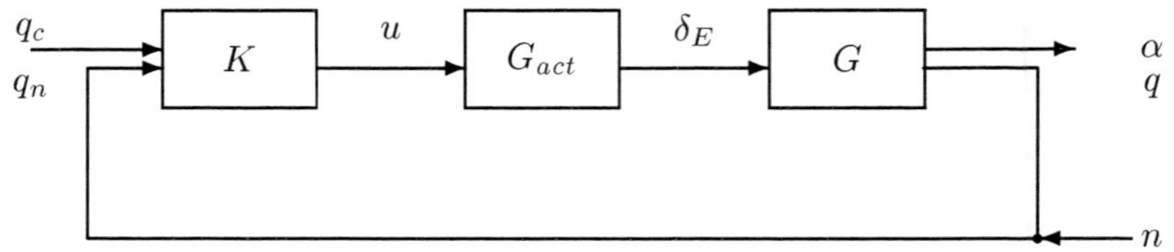


Figure 3.1: Pitch Rate Control System

The control architecture for the pitch rate command system is shown in figure 3.1. It consists of the aircraft dynamics $G$, the actuator dynamics $G_{act}$ and the controller $K$. The control structure reflects a two-degrees-of-freedom pitch rate controller. It consists of of a feedforward contribution for command shaping and a feedback contribution to reduce the effects of uncertainty on the command response. The design objectives can be divided into the following sub-objectives:

**Performance objectives**

- **Tracking performance**
  The controller should track pilot's pitch rate commands $q_c$ such that in steady state condition the system output $q$ becomes the commanded value $q_c$. We assume that the pitch rate measurement is not ideal and is corrupted with sensor noise $q_n$. The pilot's control authority is limited to a maximum pitch rate command of $q_{cmax} = 10\ [deg/s]$. The ideal dynamics $H_{id}$ of the closed-loop pitch rate response are captured in a "Handling Quality" model given by the first order transfer function:

$$H_{id}(s) = \frac{q(s)}{q_c(s)} = \frac{1.5}{s+1.5} \tag{3.2}$$

- **Angle of attack limitation**
  During a full pitch command manoeuvre the angle-of-attack is limited to $|\alpha| < 20\ [deg]$.

- **Control effort minimization**
  The actuator model for the elevator is given by:

$$G_{act}(s) = \frac{15}{s+15} \tag{3.3}$$

  Further, to prevent actuator saturation the elevator surface deflection and deflection rate are not allowed to exceed $(|\delta_E|, |\dot{\delta}_E|) < (10, 30)$ [deg, deg/s].

### Robustness Objectives

- **Parametric uncertainties**
  From robustness considerations we have to ensure that the system performs well for trim speed variations up to $\pm 10$ [m/s]. The control and stability derivatives which are the coefficients in equation 3.1 are known to have all a 20 % relative variation from their nominal trim value. Table 3.1 gives the nominal values of the control and stability derivatives and their maximum relative variations.

| Parameter | Nominal Value | Perturbation |
|---|---|---|
| $Z_\alpha$ | -0.7441 | 20 % |
| $Z_{\delta_e}$ | -0.0900 | 20 % |
| $M_{\delta_e}$ | -6.7679 | 20 % |
| $M_q$ | -1.5773 | 20 % |
| $M_\alpha$ | -1.4796 | 20 % |

Table 3.1: Stability and Control Derivatives and their Relative Variation

- **Unmodelled dynamics**
  Finally, another source of uncertainty comes from the actuator dynamics $G_{act}$ which are assumed to have high frequency unmodelled dynamics that will be defined later.

The mathematical formulation of the performance and robustness specifications is done by using using constant or frequency dependent weighting functions which are placed in the system loop upon the signals of interest. Practically, this process leads to the generalized interconnection structure $P$ from which we can develop the generalized transfer function for analysis $M$ given a controller $K$ stabilizing the closed-loop system. The weighting functions are the main design parameters and shall be altered in five successive designs in order to demonstrate how these influence the design solution. The overall control objective is to meet all the performance objectives in the face of all system uncertainties. When the specifications for the complete set of models are achieved the system is said to have *Robust Performance* which in terms of the structured singular value design means that $\mu(M)$ is less than one. The robust performance problem consists of two subproblems, the robust stability problem and the nominal performance problem. The successive steps for the design and analysis of a robust performing controller are illustrated in the following sections.

## 3.2 Setting up the design problem

### Parametric Uncertainty Modelling

The uncertainty modelling procedure is illustrated on the aircraft problem. Due to changing operating conditions the state space entries of the nominal model (equation 3.1) vary substantially.

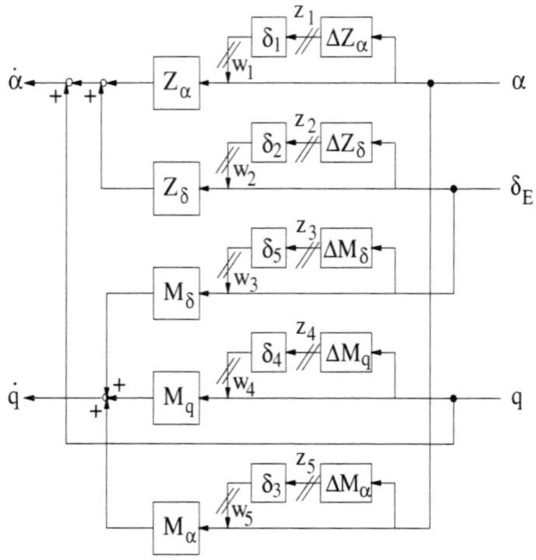


Figure 3.2: Block diagram of example system with perturbations

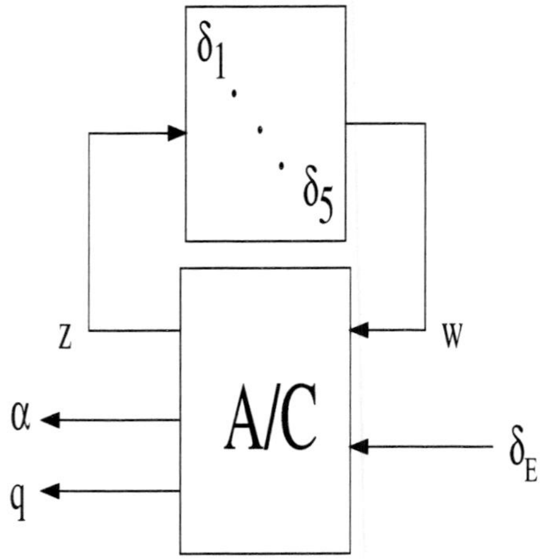


Figure 3.3: Representation of example system in LFT-form

The first step in the parametric uncertainty modelling is to draw the system dynamics given by equation 3.1. Then, we place on each stability and control derivative the relative variation given in table 3.1. This process leads to the uncertain system drawn figure 3.2. The perturbations in the table are the maximum absolute changes the parameters can undergo. These perturbations are normalized by introducing scalings.
We may change any of these $\delta$'s arbitrarily within the given bounds. The model parametrized in the $\delta$'s reflects a *set of models*.

From this uncertain model representation we need to derive an LFT representation as shown in figure 3.3 to reflect how the uncertainties affect simultaneously the model. It consists of an invariant part of the model $A/C$ and the uncertain elements where the delta's stacked in the matrix $\Delta$. This process is known as "pulling out the delta's". All uncertain elements $\delta_1$ ... $\delta_5$ are diagonally augmented in the perturbation matrix $\Delta = diag(\delta_1 \; ... \; \delta_5)$. In figure 3.2 the signals to and from the delta's have been cut; the signals $(z_1...z_5)$ become the outputs from the constant part A/C and inputs of $\Delta$, while the opposite holds for the signals $(w_1...w_5)$. We can now read off all signal relations given by the mapping and build the following matrix:

$$\left[\begin{array}{c} \dot{q} \\ \dot{\alpha} \\ \hline z_1 \\ z_2 \\ z_3 \\ z_4 \\ z_5 \\ \hline q \\ \alpha \end{array}\right] = \left[\begin{array}{cc|ccccc|c} M_q & M_\alpha & 0 & 0 & M_{\delta_e} & M_q & M_\alpha & M_{\delta_e} \\ 1 & Z_\alpha & Z_\alpha & Z_{\delta_e} & 0 & 0 & 0 & Z_{\delta_e} \\ \hline \Delta Z_\alpha & 0 & 0 & 0 & 0 & 0 & 0 & 0 \\ 0 & 0 & 0 & 0 & 0 & 0 & 0 & \Delta Z_\delta \\ 0 & 0 & 0 & 0 & 0 & 0 & 0 & \Delta M_\delta \\ 0 & \Delta M_q & 0 & 0 & 0 & 0 & 0 & 0 \\ \Delta M_\alpha & 0 & 0 & 0 & 0 & 0 & 0 & 0 \\ \hline 1 & 0 & 0 & 0 & 0 & 0 & 0 & 0 \\ 0 & 1 & 0 & 0 & 0 & 0 & 0 & 0 \end{array}\right] \left[\begin{array}{c} q \\ \alpha \\ \hline w_1 \\ w_2 \\ w_3 \\ w_4 \\ w_5 \\ \hline \delta_E \end{array}\right] \qquad (3.4)$$

The so obtained matrix $A/C$ closed with the block diagonal matrix $\Delta$ as shown in figure 3.3 provides the required LFT formulation of the uncertain aircraft dynamics covering all possible parameter variations. Notice that the parametric uncertainty modelling process reveals that uncertainty that is unstructured at parameter level (component level) becomes structured at system level ($\mathbf{A/C}$).

### Unmodelled Dynamics

Another possible way to capture a model set is shown for the actuator case. The elevator position is changed via an actuator having first order dynamics $G_{act}(s) = \frac{15}{s+15}$. The device is assumed to give position errors up to 20% in a frequency range up to 1 $rad/s$, while at higher frequencies this may be even more. The variation in the position error along the frequency is represented by a first order transfer function $W_{pert}(s) = 0.20\frac{s/1+1}{s/40+1}$. The model set covered by the uncertain actuator dynamics is given by

$$\tilde{G}_{act} = \{G_{act}\,(1 + \delta_6 W_{pert}(s)) : W_{pert}(s) \text{ stable and } \|\delta_6\ \|_\infty \leq 1\}.$$

The weighting function is used to normalize the unknown perturbation $\delta_6$ and at any frequency $\omega$ the magnitude of $W_{pert}$ represents the relative uncertainty level in the actuator model. This type of modelling is called multiplicative uncertainty modelling.

### Placing the Performance Objectives

Upon the chosen control architecture given in figure 3.1 we place at the physical location in the system, namely at signal level our requirements. The design requirements are translated into weighting functions chosen either frequency dependent or constant to bound the signals of interest.

In doing so we end up with the interconnected system depicted in figure 3.4. This system is called the *general interconnection structure*. It consists of the aircraft model $G(s)$ with its parametric uncertainties, the uncertain actuator $\tilde{G}_{act}(s)$, the controller $K(s)$ to be designed, an ideal model $H_{id}(s)$ which we want to match and the performance weighting filters that place emphasis on the frequency content and amplitude on the signals of interest. The interconnection structure in figure 3.4 is the pictorial equivalent of the connection between the generalized plant $P$ together with the stabilizing controller $K$ and the uncertainty $\Delta$.

When dealing with optimization techniques it is important to scale the sys-

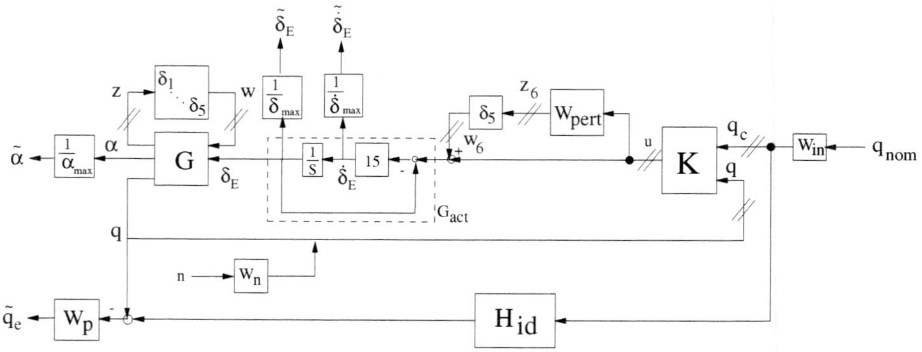


Figure 3.4: Interconnection structure of the example system

tem's units appropriately. The pitch rate command signal is normalized with respect to the maximum expected commands with the filter $W_{in}$. The pitch rate command input $q_c$ goes through the ideal model. The difference between the ideal model response and the actual pitch rate measurement $q$ is the *tracking error*. To emphasize how large and up to what frequency the error reduction should last, a filter $W_p(s)$ reflecting the tracking objective is placed on the error signal.

Tracking should not be achieved at the cost of too high control activity. To prevent this the elevator deflection and deflection rate are penalized. The deflection and deflection rate are weighted by the inverse of their maximum allowed values $W_{\delta_E} = 1/\delta_{E_{max}}$ and $W_{\dot{\delta}_E} = 1/\dot{\delta}_{E_{max}}$.

When one of these weighted signals is larger than one, in the $H_\infty$ norm sense, the individual performance objective is violated.

The angle-of-attack is a free variable in the design. To prevent the aircraft from stalling during a full pitch rate command an angle-of-attack limiting function is necessary. This is introduced in the design by the choice of a performance specification $W_\alpha$ penalizing the signal $\alpha$. It is chosen as the inverse of the maximum allowable angle-of-attack as $W_\alpha = 1/\alpha_{max}$. In the case the norm of the weighted angle-of-attack $W_\alpha$, becomes larger than one, the stall limit is violated.

To reclect imperfections in the pitch rate measurement signal a filter $W_n$ coloring the noise $n$ is added to the pitch rate measurement signal as depicted in figure 3.4. The filter $W_n$ scales the normalized measurement noise $n$ as a function of the frequency.

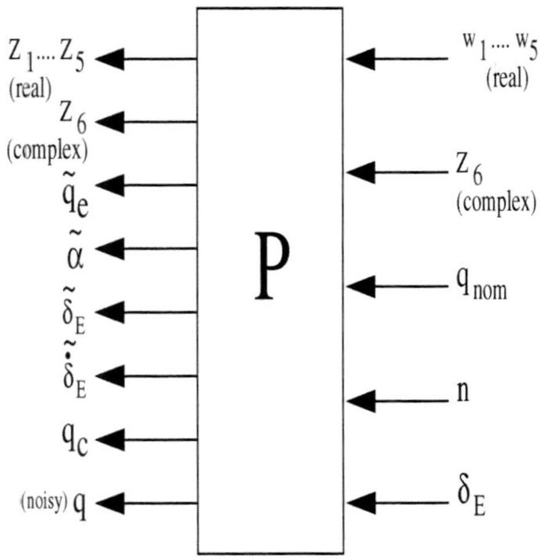


Figure 3.5: Open-loop interconnection structure

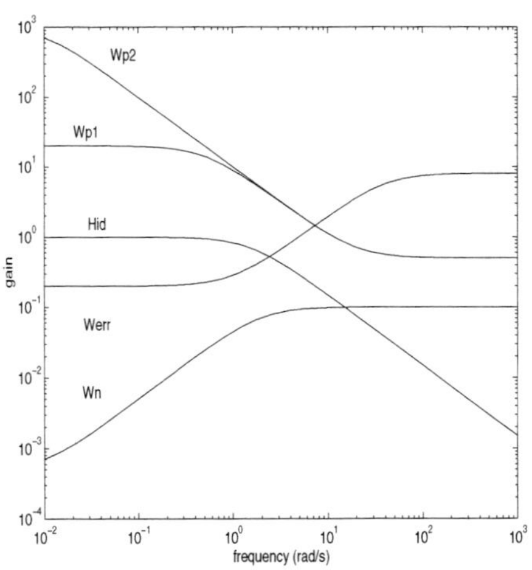


Figure 3.6: Weighting functions for design example (not scaled with $W_{in}$)

## 3.3 Weighting Function Selection

When disconnecting in figure 3.4 the controller $K$ and all the uncertainties $\Delta$ we can read off the signal relations that allow us to deterimine the open-loop interconnection structure $P$ shown in figure 3.5. The open-loop interconnection system $P$ has three pairs of inputs and outputs organized accordingly. The $\tilde{\cdot}$ on the signals is used to indicate that these signals are weighted.
The first pair of signals corresponds to the uncertainty channels ($z$ $w$), the second pair to the performance channels given by $e = [\tilde{q}_e,\ \tilde{\alpha},\ \tilde{\delta}_E,\ \tilde{\dot{\delta}}_E]$, $d = [q_{nom},\ n]$ and the third pair to the measurement/control channels denoted with $y = [q_c,\ q + W_n n]$, and $u = [\delta_E]$.

The general open-loop interconnection $P$ contains all the problem data required for design including the weighting functions. Up to now, a choice has been made on the the controller architecture, the set of models, and the performance signals we wish to take into account.

The weighting functions are our design parameters. The choice of the weighting functions is based on an understanding of how the various conflicting objectives interact with each other in a closed-loop configuration. The primary analysis is based on the analytical expression for the generalized closed-loop system $M = F_l(P, K)$ for which is assumed a stabilizing controller $K$ is given.

The question arising now is, how do the weights have to look like if we want our closed-loop system to achieve robust performance. The following steps are meant to illustrate the rationale behind this weighting function selection process.

### Step 1: Deriving the General Interconnection System $P$

For simplicity, we disregard in the aircraft model the parametric uncertainties $\delta_R = diag(\delta_1, \ldots, \delta_5) \in \Delta_{\mathbf{R}}$ and consider only the complex uncertainty $\delta_6 \in \Delta_{\mathbf{C}}$ at the actuator.

Recall, the first step is to determine graphically from figure 3.4 the transfer functions that constitute the general open-loop interconnection structure $P$ as shown in figure 3.5. The analytical expression for $P$ is obtained by reading-off the the remaining loop equations in figure 3.4 when both, the controller $K$ and the uncertainty $\Delta$ are disconnected from the weighted closed-loop structure. The signal locations at which the loop has been disconnected are shown in figure 3.4 with two diagonal lines.

Denote with $[G_q, G_\alpha]$ the transfer function from $\delta_E$ to $(q, \alpha)$. The $3 \times 3$ general open-loop interconnection structure given by $P$:

$$\left[\begin{array}{c} z \\ \hline e \\ \hline y \end{array}\right] = \left[\begin{array}{c|c|c} P_{11} & P_{12} & P_{13} \\ \hline P_{21} & P_{22} & P_{23} \\ \hline P_{31} & P_{32} & P_{33} \end{array}\right] \left[\begin{array}{c} w \\ \hline d \\ \hline u \end{array}\right] \tag{3.5}$$

becomes in our case the system shown in figure 3.5 by reading-off the loop equations from figure 3.4 for which the analytical expression is given by:

$$\left[\begin{array}{c} z \\ \hline \tilde{q}_e \\ \tilde{\alpha} \\ \tilde{\delta}_E \\ \tilde{\dot{\delta}}_E \\ \hline q_c \\ q \end{array}\right] = \left[\begin{array}{c|cc|c} 0 & 0 & 0 & W_{pert} \\ \hline -W_p G_q G_{act} & W_p H_{id} W_{in} & 0 & -W_p G_q G_{act} \\ W_\alpha G_\alpha G_{act} & 0 & 0 & W_\alpha G_\alpha G_{act} \\ 0 & 0 & 0 & W_{\delta_E} G_{act} \\ 0 & 0 & 0 & W_{\dot{\delta}_E} s G_{act} \\ \hline 0 & W_{in} & 0 & 0 \\ G_q G_{act} & 0 & W_n & G_q G_{act} \end{array}\right] \left[\begin{array}{c} w \\ \hline q_{nom} \\ n \\ \hline \delta_E \end{array}\right] \tag{3.6}$$

### Step 2: Deriving the General Interconnection System for Analysis $M$

Given the controller $K$ and the transfer fuction $P$ one can determine the analytical expression of the closed-loop system $M = F_l(P, K)$ for analysis via the following shortcut:

$$M_{ij} = P_{ij} + P_{i3} \left[I - K P_{33}\right]^{-1} K P_{3j} \quad i,\ j = 1,\ 2$$

The controler $K$ is partitioned accordingly into $[K_c \ \ K_f]$, here $K_c$ represents the *command part* from command $q_c$ to $u$, and $K_f$ reflects the *feedback task* from measurement $q$ to $u$.

The Input Sensitivity function $S_i$ is given by:

$$\begin{aligned} [I - KP_{33}]^{-1} &= \left[I - [K_c \; K_f] \; [0 \; G_q G_{act}]^T\right]^{-1} \\ &= [I - K_f G_q G_{act}]^{-1} \\ &= S_i \end{aligned}$$

so that with the Input Complementary Sensitivity $T_i$ we get $S_i + T_i = I$.

The Robust Stability Block $M_{11}$ becomes:

$$\begin{aligned} M_{11} &= 0 + W_{pert} S_i [K_c \; K_f] \; [0 \; G_q G_{act}]^T \\ &= W_{pert} S_i K_f G_q G_{act} \\ &= W_{pert} T_i \end{aligned}$$

The Interaction block $M_{12}$:

$$\begin{aligned} M_{12} &= W_{pert} S_i [K_c \; K_f] \begin{bmatrix} W_{in} & 0 \\ 0 & W_n \end{bmatrix} \\ &= [W_{pert} S_i K_c W_{in} \quad W_{pert} S_i K_f W_n] \end{aligned}$$

and $M_{21}$:

$$\begin{aligned} M_{21} &= \begin{bmatrix} -W_p G_q G_{act} \\ W_\alpha G_\alpha G_{act} \\ 0 \\ 0 \end{bmatrix} + \begin{bmatrix} -W_p G_q G_{act} \\ W_\alpha G_\alpha G_{act} \\ W_{\delta_E} G_{act} \\ W_{\dot{\delta}_E} s G_{act} \end{bmatrix} S_i K_f G_q G_{act} \\ &= \begin{bmatrix} -W_p G_q G_{act} S_i \\ W_\alpha G_\alpha G_{act} S_i \\ W_{\delta_E} G_{act} T_i \\ W_{\dot{\delta}_E} s G_{act} T_i \end{bmatrix} \end{aligned}$$

Finally, the Nominal Performance block $M_{22}$ is:

$$\begin{aligned} M_{22} &= \begin{bmatrix} W_p H_{id} W_{in} & 0 \\ 0 & 0 \\ 0 & 0 \\ 0 & 0 \end{bmatrix} + \begin{bmatrix} -W_p G_q G_{act} \\ W_\alpha G_\alpha G_{act} \\ W_{\delta_E} G_{act} \\ W_{\dot{\delta}_E} s G_{act} \end{bmatrix} S_i \begin{bmatrix} K_c W_{in} & K_f W_n \end{bmatrix} \\ &= \begin{bmatrix} W_p (H_{id} - G_q G_{act} S_i K_c) W_{in} & -W_p G_q G_{act} S_i K_f W_n \\ W_\alpha G_\alpha G_{act} S_i K_c W_{in} & W_\alpha G_\alpha G_{act} S_i K_f W_n \\ W_{\delta_E} G_{act} S_i K_c W_{in} & W_{\delta_E} G_{act} S_i K_f W_{in} \\ W_{\dot{\delta}_E} s G_{act} S_i K_c W_{in} & W_{\dot{\delta}_E} s G_{act} S_i K_f W_n \end{bmatrix} \end{aligned}$$

The complete analysis system for Robust Performance is $M$:

$$M = \text{diag} \left[ \begin{array}{c} W_{pert} \\ \hline W_p \\ W_\alpha \\ W_{\delta_E} \\ W_{\dot{\delta}_E} \end{array} \right]^T \left[ \begin{array}{c|cc} T_i & S_i K_c & S_i K_f \\ \hline -G_q G_{act} S_i & H_{id} - G_q G_{act} S_i K_c & -G_q G_{act} S_i K_f \\ G_\alpha G_{act} S_i & G_\alpha G_{act} S_i K_c & G_\alpha G_{act} S_i K_f \\ G_{act} T_i & G_{act} S_i K_c & G_{act} S_i K_f \\ s G_{act} T_i & s G_{act} S_i K_c & s G_{act} S_i K_f \end{array} \right] \text{diag} \left[ \begin{array}{c} I \\ \hline W_{in} \\ W_n \end{array} \right]^T \tag{3.7}$$

### Step 3: Robust Stability

Applying the General Analysis Theorem 2.4.2, the posed flight control problem is solved when robust performance is achieved, that is when the structured singular value $\mu$ of the general analysis system $M$ is less than unity over all frequencies $\omega$.

According to the various block definitions the system is robustly stable for all allowable perturbations $\Delta$ if and only if

$$||W_{pert}T_i||_\infty < 1$$

To achieve the robust stability condition for the $M_{11}$ block the weight $W_{pert}$ is chosen in a way to force the loop gain to roll off-at higher frequencies while at low frequencies where the magnitude of the input complementary sensitivity function $|T_i| \approx 1$ we choose the magnitude of $W_{pert}$ to reflect 20% of relative uncertainty.

At higher frequencies we have $|T_i| \approx |K_f G|$, while the uncertainty $|W_{pert}|$ still increases to account for unmodeled dynamics. The weighted Loop Gain needs to become small satisfying

$$|W_{pert}K_f G_q G_{act}| < 1$$

therefore, in order to achieve robust stability, we need that the unweighted Loop Gain does not cross the boundary given by:

$$|K_f G_q G_{act}| < 1/|W_{pert}|$$

Notice, that the constraint given by the weight $W_{pert}$ will have a minor effect on the Loop Gain at higher frequencies. This is because the plant together with the actuator have already sufficient gain roll-off in this frequency range.

The weight is however of crucial importance in the cross-over frequency of $W_{pert}$ which is used as a design parameter to limit the bandwidth of the imput complementary sensitivity function $T_i$. Of course we would like to maximize this frequency but it should always be limited below the actuator bandwith in order to prevent excitation of structural modes by high frequency control signals.

### Step 4: Performance Analysis

Nominal performance is achieved if and only if:

$$\left\| \text{diag}\begin{bmatrix} W_p \\ W_\alpha \\ W_{\delta_E} \\ W_{\dot{\delta}_E} \end{bmatrix}^T \begin{bmatrix} H_{id} - G_q G_{act} S_i K_c & -G_q G_{act} S_i K_f \\ G_\alpha G_{act} S_i K_c & G_\alpha G_{act} S_i K_f \\ G_{act} S_i K_c & G_{act} S_i K_f \\ s G_{act} S_i K_c & s G_{act} S_i K_f \end{bmatrix} \text{diag}\begin{bmatrix} W_{in} \\ W_n \end{bmatrix}^T \right\|_\infty < 1$$

while robust performance is achieved if and only if:

$$\mu(M) < 1 \;\forall\, \omega$$

with $M$ given by 3.7.

Next, are discussed the various contributions that constitute the nominal performance objective given the block $M_{22}$.

### Tracking Objective

To achieve the nominal tracking objective, the following constraint needs to hold:

$$||W_p(H_{id} - G_q G_{act} S_i K_c) W_{in}||_\infty < 1$$

At low frequencies the ideal model has unit gain, $|H_{id}| = 1$. Assume further that the command part of the controller behaves as $K_c \approx -K_f$. This means that the previous constraint reads as:

$$|W_p(I + G_q G_{act} S_i K_f) W_{in}| = |W_p(I - G_q G_{act} K_f)^{-1} W_{in}| < 1$$

At low frequencies when the Loop Gain is high, we have approximately:

$$|W_p(G_q G_{act} K_f)^{-1} W_{in}| < 1$$

so that at low frequencies Loop Gain is bounded by:

$$|G_q G_{act} K_f| > |W_p W_{in}|$$

We choose the input weight $W_{in}$ constant over the lower frequency range. Furthemore the transfer function $G_q G_{act}$ has a constant gain at low frequencies. This means that we can mainly influence the low frequency shape of the controller $K$ via a proper shape of $W_p$. If the tracking error level at DC is not sufficiently small we can, by increasing the low frequency gain of $W_p$, force the controller optimization algorithm to build more integral action into $K$.

Observe that the tracking performance may be destroyed by the contribution of the noise given by:

$$-W_p G_q G_{act} S_i K_f W_n$$

Since $|G_q G_{act} S_i K_f| \approx 1$ at low frequencies, the magnitude of the weights need at least to satisfy:

$$|W_p W_n| < 1$$

But, if the weighting $W_p$ is chosen to have too high gain at low frequencies to increase the controller gain, the previous requirement is easily violated.

There are two solutions to overcome this problem. In first instance we can make the noise input low where the magnitude of the performance weight $W_p$

is high. The constraint is satisfied as long as we choose the magnitude of the noise weight $W_n$ needs to have approximately the inverse shape of $W_p$.

The second solution is to feed the noise signal to the performance filter, such that the expression

$$H_{id}q_c - (q + W_n n)$$

becomes weighted. In this way the steady-state value (at $\omega = 0$) of $W_n$ accounts for example for a sensor bias. This "bias" may violate the performance index, because this is based on the error between the reference and the exact output.

Applying the second solution, the error is related to the same biased measurement as the controller receives. Remark that in a standard feedback configuration the transfer function of the noise to the output is changed from a complementary sensitivity function to a sensitivity function, which has low gain at low frequencies. We choose here to lower the gain of $W_n$ if necessary; as will be shown later this has also an advantage when designing a nominal controller for the plant, that is when all uncertainties are set to zero.

### Control Effort Minimization

The control effort requirement involves the transfer functions $W_{\dot{\delta}_E} sG_{act}S_iK_cW_{in}$ and $W_{\dot{\delta}_E} sG_{act}S_iK_f$ that mainly play an important role at higher frequencies.

In this frequency range the input sensitivity function is of magnitude $|S_i| \approx 1$, while the commanded elevator deflection rate $sG_{act} = 15j\omega/j\omega + 15 \approx 15$ as $s \to \infty$.
This means that the controller $K$ must fulfil at least the following two constraints:

$$|K_c| < 1/|W_{\dot{\delta}_E} 15W_{in}|$$

and

$$|K_f| < 1/|W_{\dot{\delta}_E} 15W_n|$$

respectively.
These weightings impose an upperbound on the high frequency gain of the controller. In many cases the controller has already enough rolls-off at higher frequencies. This means that in this frequency region these weights have no substantial effect on the design. However, in the presence of plant perturbations or severe disturbances it is important to penalize the actuator rates. This is done to prevent the controller $K$ from producing control signals with deflection rates beyond the physical limits of the plant to cause a severe rate saturation.

## 3.4 Quantitative Weighting Function Selection

The maximum pitch rate commands ($q_c$) are given to be up to 10 degrees per second. To normalize the input commands to one the input filter $W_{in}$ is chosen to be $W_{in} = 10/57.3\ rad/s$.

Next, to illustrate how the the controller shape can be influenced two performance weights $W_{p1}$ and $W_{p2}$ are chosen:

$$W_{p1} = \frac{20}{|W_{in}|}\frac{s/20+1}{s/0.5+1} \quad W_{p2} = \frac{1000}{|W_{in}|}\frac{s/20+1}{s/0.01+1} \tag{3.8}$$

The cross-over frequencie $\omega_c$ of both filters $W_{p1}$ and $W_{p2}$ has been chosen the same at $\omega_c = 20$ [rad/sec]. The choice of this cross-over frequency is an important design consideration to reflects the decay rate at which the tracking errors need to be reduced. In the low and the mid-frequency range there are two relevant parameters for shaping the filter $W_p$. This is the steady-state gain and the cross-over frequency of the filter $W_p$. The steady-state gain of this filter determines actually the inverse of normalized tracking error level. In the case of $W_{p1}$ the DC-gain is 20 imposing a 5% tracking error to unit commands. In the second case $W_{p2}$ the requirements becomes more stringent imposing a 0.1% tracking error. These errors must be kept for frequencies ranging from DC up to the cross-over frequency of the filter. It is advisable during a design to alter only one of these parameters at-a-time since both heavily affect the performance requirement.

To be more specific suppose that after controller design a time domain simulation reveals the the steady-state error appears to be too large. We can reduce this error by simply increasing the gain of weighting filter. But this means that also the cross-over frequency will automatically increase, leading to unintended other effects due to the faster decay rate requirement. Therefore, it is better to fix the frequency at which a good cross-over point is found, and try to extend the slope of the weight into the low frequency range in order to increase the low frequency gain to enforce integral control action. The described effect for the filters $W_{p1}$ and $W_{p2}$ can be seen in figure 3.6. Notice also, that both weights flatten out at $\omega = 10\ rad/s$, which is below the actuator bandwidth. Keeping enough frequency separation between the tracking decay and the actuator bandwidth is an efficient way to limit excessive transient behaviour.

The weights on the elevator deflection and deflection rate are chosen as:

$$W_{\delta_E} = \frac{1}{\delta_{E_{max}}} = \frac{1}{|W_{in}|}\ rad^{-1} \quad W_{\dot{\delta}_E} = \frac{1}{\dot{\delta}_{E_{max}}} = \frac{1}{3|W_{in}|}\ (rad/s)^{-1}$$

To limit the angle-of-attack we choose a constant weighting filter:

$$W_\alpha = \frac{1}{\alpha_{max}} = \frac{1}{2|W_{in}|} = \frac{57.3}{20}$$

Finally, the chosen noise filter is frequency dependent and defined as:

$$W_n = 0.0005|W_{in}|\frac{s/0.01+1}{s/2+1} = \frac{0.01}{57.3}\frac{s/0.01+1}{s/2+1}$$

The DC gain is chosen low taking into account that we also have to satisfy the performance index imposed by $W_{p1}$ or $W_{p2}$. Remark, that all weighting functions are expressed in terms of $W_{in}$ relating in this way all objectives to the normalized maximum pitch rate command signal $q_c$.

## 3.5 Control Synthesis

In order to reflect the fundamental trade-off between the achievable performance and robustness we shall illustrate the effects of various weighting function selections for as well the performance as the uncertainty by studying the following design cases:

1. **Nominal Performance Design:** All uncertainties are set to zero and we designate the controller resulting from this synthesis with $K_{nom}$.
2. **Complex Uncertainty Design:** We assume the plant has only a complex uncertainty $\delta_6 \in \Delta_\mathbf{c}$ at the actuators. The resulting controller will be denoted as $K_2$. For this uncertainty we shall design two controllers $K_2(W_{p1})$ and $K_2(W_{p2})$ with increasing tracking performance objective respectively $W_{p1}$ and $W_{p2}$. The augmented perturbation structure related to the robust performance index is given by $\Delta_2 = diag(\delta_6, \Delta_{perf})$.
3. **Real and Complex Uncertainty Design:** Taking all uncertainties into account leads in the synthesis process to controller $K_3$. The real uncertainty $\delta_R = diag(\delta_1, \ldots, \delta_5) \in \Delta_\mathbf{R}$ corresponds to the potential variations in the control and stability derivatives. For the combined real and complex uncertainty we shall design two controllers $K_3(W_{p1})$ and $K_3(W_{p2})$ with increasing tracking performance objective respectively $W_{p1}$ and $W_{p2}$. The augmented perturbation structure related to the robust performance objective is denoted with $\Delta_3 = diag(\delta_1, \ldots, \delta_5, \delta_6, \Delta_{perf})$.

| | $K_{nom}$ | $K_2(W_{p1})$ | $K_2(W_{p2})$ | $K_3(W_{p1})$ | $K_3(W_{p2})$ |
|---|---|---|---|---|---|
| $\bar{\sigma}(M)$ | 0.89 | 28.66 | 28.66 | 28.66 | 28.66 |
| $\mu(M)$ | | 1.71 | 1.86 | 1.88 | 2.11 |
| $\mu(M)$ | | 1.51 | 1.60 | 1.78 | 1.89 |
| order | 6 | 9 | 9 | 15 | 15 |

Table 3.2: Achieved robust performance levels

In table 3.2 the achieved robust performance levels expressed in $\mu$ values for all configurations are summarized. Row # 1 shows the results of the pure $H_\infty$ optimization. Rows # 2 and # 3 reveal the robust performance levels achieved after the first and second $D - K$ iteration. In the last row the order of the resulting controllers is given. The table reflects a well known fact that the robust performance level decreases as the uncertainty and performance levels increase.

## 3.6 Analysis of the Results

### Analysis: Shape of the Controller

In figure 3.7 the frequency responses of the controllers are depicted. As already mentioned in section 3.3, the controller will not necessarily roll-off at higher frequencies, since the combination of the actuator and the plant model already shows this behaviour. We can see that controller shapes flatten out at higher frequencies and lower gains. In the section 3.3 we have seen that the gain of $K_c$ at higher frequencies has to satisfy: $|K_c| < 1/|W_{\dot{\delta}_E} 15 W_{in}| = 1/(57.3/30 * 15 * 10/57.3) = 0.2$. This is confirmed by figure 3.7. For $K_f$ we have: $|K_f| < 1/|W_{\dot{\delta}_E} 15 W_n| = 1/(57.3/30 * 15 * 0.1 * 10/57.3) = 2$. This is also satisfied.

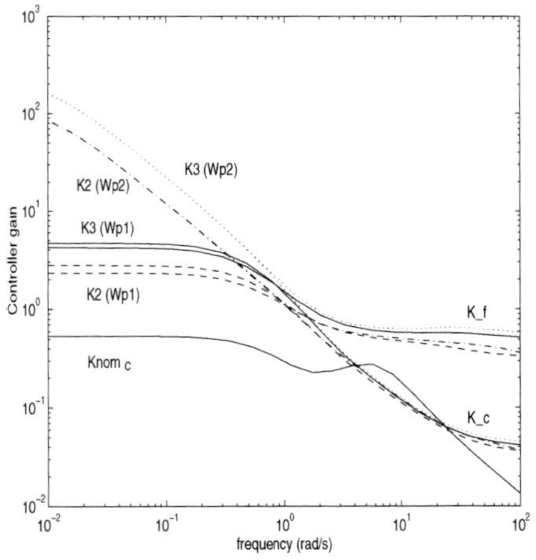


Figure 3.7: Controller frequency responses

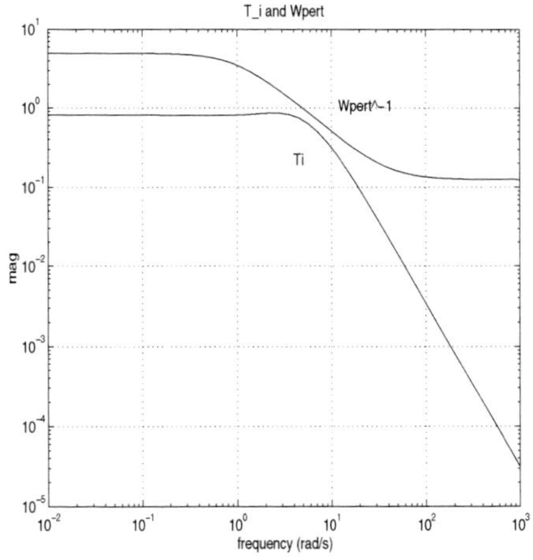


Figure 3.8: Input compl. sensitivity function with $W_{pert}^{-1}$

Due to the actuator uncertainty the controller will limit its bandwidth at frequencies where the uncertainty starts to become important. This is illustrated by the fact that the bandwidth of the input complementary sensitivity function $T_i$ is limited by the filter $W_{pert}$. The frequency response plots of $T_i$ and $W_{pert}^{-1}$ are given in figure 3.8. Indeed $T_i$ rolls-off near the cross-over frequency of $W_{pert}$. In this way we prevent unmodelled higher order dynamics from excitation by keeping the control actions within the lower frequency range. The controller $K_{nom}$ for the nominal plant is shown in figure 3.7, $K_{nom}$ has only a feedforward action $K_c$ and almost no feedback since $K_f \approx 0$ at each frequency. The feedforward action $K_c$ approximately inverts the plant to deliver the ideal model behavior. We already knew that in absence of uncertainty there is no necessity for feedback. Interestingly, the method delivers automatically the required system property.

Another interesting effect is the influence of the performance weight $W_p$. By increasing the slope into the low frequency range ($W_{p1} \rightarrow W_{p2}$) the controller gain does exactly the same. In this way we force the optimization algorithm to build integration or even double integration by taking an 40 $dB$ slope on the weighting function. As in classical Bode loop shape techniques, the internal model principle holds and to follow step commands we need integral action. By increasing the tracking error requirements imposes increasing low frequency control gain.

### Analysis: Nominal Performance $\bar{\sigma}(M_{22})$

Beside the shape of the controller we are interested in the trade-offs made. In figure 3.9 we have plotted for all controllers the nominal performance level $\bar{\sigma}(M_{22})$. The overall shape is, a high value of $\bar{\sigma}(M_{22})$ at low frequency corresponding to an effective tracking requirement at these frequencies. At higher frequencies there is no performance requirement so that the plot of $\bar{\sigma}(M_{22})$ indeed rolls-off.

When comparing all controllers in figure 3.9 the best nominal performance level is achieved by $K_{nom}$ for the system without uncertainty since $\bar{\sigma}(M_{22}) \approx 0.9$. For the other controllers the performance level is higher (worse). We know that the complex uncertainty is about 20 % in the low frequency range. This is, for the system with only complex uncertainties about the level of the performance degradation with respect to the nominal system.

It is surprising that the third controller for the most uncertain plant achieves a better nominal performance level than the second controller. The conclusion is that the real uncertainty at low frequencies, offsets the effect of the complex uncertainty with respect to the nominal performance. Notice that this effect is reversed at high frequencies.

For comparison we add the nominal performance plot for $K_2$ designed with $W_{p_2}$. Note the tighter weight $W_{p2}$ with respect to $W_{p1}$ induces only a modest performance degradation for the $K_2$ design.
It is interesting to plot the unweighted closed-loop transfer functions together with the penalizing weighting functions. When the requirement is achieved these transfer functions lie below the inverse of their weights.

As an example we consider the $K_2$ design. In the figures 3.11, 3.12, 3.13 and 3.14 the unweighted transfer functions from $q_{nom}$ to $q_e$, $\alpha$, $\delta_E$ and $\dot{\delta}_E$ are shown with the inverses of the weighting functions previously placed on those signals. The first figure shows the effect of increasing the tracking objective from $W_{p1}$ to $W_{p_2}$. In both case the tracking objective is met and we see that $W_{p2}$ has pushed the transfer function $q_e/q_{nom}$ down in the low frequency range by building integral

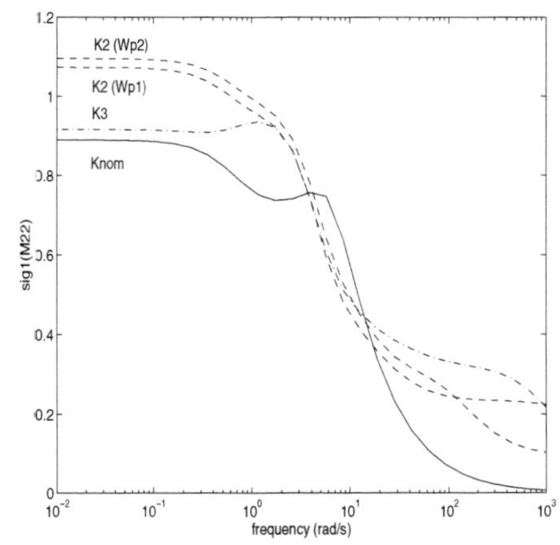


Figure 3.9: Nominal performance levels controllers

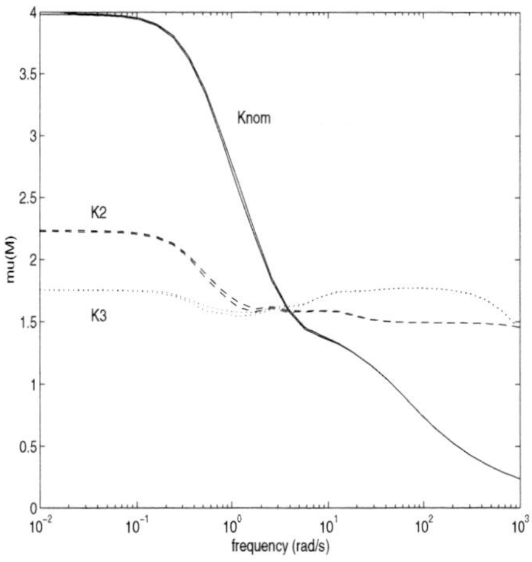


Figure 3.10: Robust performance levels controllers

action in the controller. The elevator and angle-of-attack command responses are of importance at low frequencies and respect the limits imposed by the weights. The elevator deflection rate requirement is also satisfied but plays an important role at higher frequencies.

In the same way we can look at the influence of noises to the performance outputs. If the nominal performance requirement is achieved, all unweighted transfer functions in $M_{22}$ lie below the inverses of their weights. This does not hold the other way around. Separate input signals do not necessarily need to violate a certain bound on an output channel but their combination possibly can do that. Therefore it is often important to watch the interaction channels given by $M_{12}$ and $M_{21}$.

### Analysis: Robust Performance $\mu_{\Delta_3}(M)$

The actual robust performance levels for each design are summarized in table 3.2.

For the comparison of the achieved off-design robust performance levels a $\mu$-test over largest uncertainty $\Delta_3$ is applied on all achieved controllers.

The resulting robust performance levels are shown in figure 3.10. None of the controller achieves robust performance. By rescaling the performance channels with the inverse of the worst case performance level robust performance i.e. $\mu(M) < 1$ can be achieved provided robust stability is achieved, i.e. $\mu(M_{11}) < 1$. In this way a new controller can be redesigned for less tight performance objectives but with the guarantee that $\mu(M) < 1$.

Our purpose here is more to reveal the effects of off-design specifications for the controllers $K_2$ and $K_{nom}$. The question we have in mind is, how does the

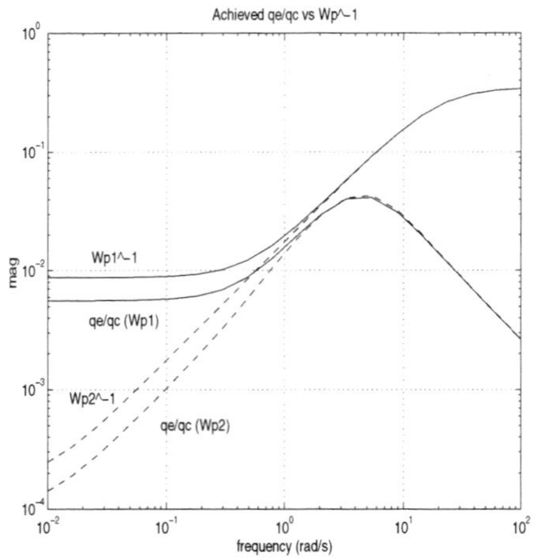


Figure 3.11: $|\frac{q_e}{q_{nom}}|$ with inverses of weights

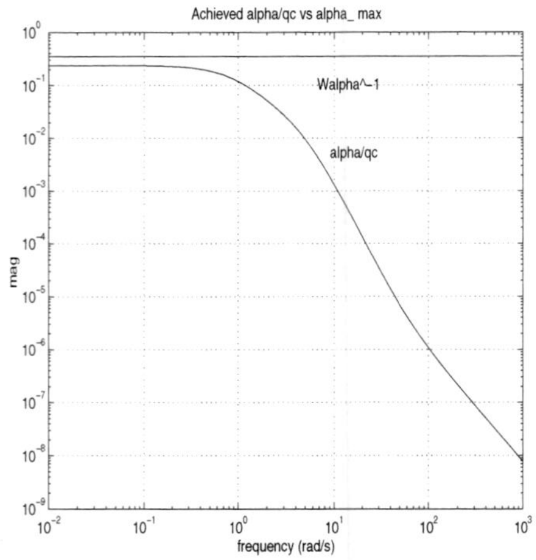


Figure 3.12: $|\frac{\alpha}{q_{nom}}|$ with inverses of weights

performance degrade of a designed controller in the face of uncertainties that have not been taken into account by design?

### The Nonimal Controller $K_{nom}$

In figure 3.10 the nominal controller $K_{nom}(W_{p1}))$ performs worst with a 400 % robust performance degradation at low frequencies to complex and real uncertainties. At higher frequencies it has a better robust performance level than $K_2$ and $K_3$. The result is expectable, since $K_{nom}$ is pure feedforward controller.

### The Controllers $K_2$ and $K_3$

Figure 3.10 further shows the $\mu$-plots of the $K_2(W_{p1})$ and $K_3(W_{p1})$ design which is the medium performance case with weight $W_{p1}$. The high performance case for $W_{p2}$ is not shown in the figure since the performance collapse of $K_{nom}(W_{p2})$ would have affected the scale of the plots.

The $K_2$ controller shows a 70 % performance degradation due to the real perturbation offset. In the $K_3$ design the real perturbations have been assumed complex which is a conservatism of the $\mu$-synthesis. Compared to the $K_2$ case the offset analysis of the $K_3$ design shows an improved robust performance level in the low frequency range at the cost of robust performance degradation at higher frequencies. The shape of the robust performance level in the $K_3$ case has flattend out over the whole frequency range. A better balance between the performance and robustness requirements has been achieved. This effect is caused by the fact that in the design we have taken into account the uncertainty in the stability and control derivatives which prioritizes the robust stability re-

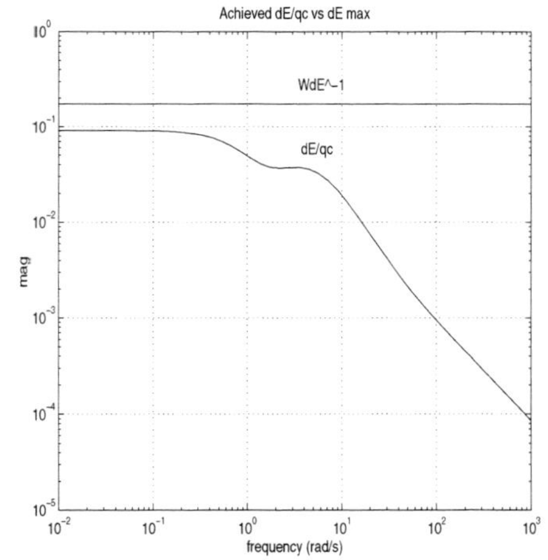


Figure 3.13: $|\frac{\delta_E}{q_{nom}}|$ with inverses of weights

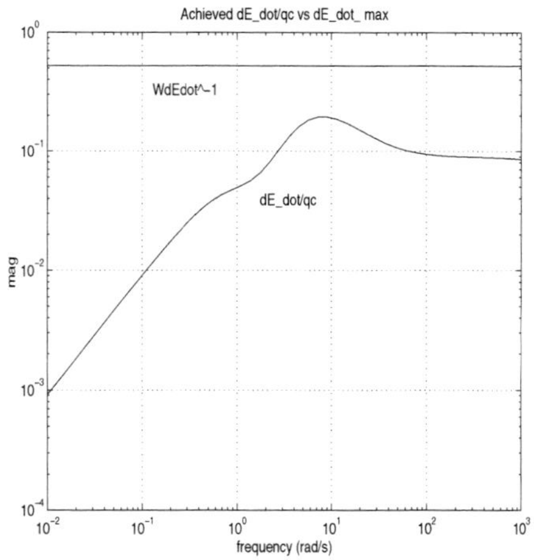


Figure 3.14: $|\frac{\dot{\delta}_E}{q_{nom}}|$ with inverses of weights

quirement.

**Analysis: Robust Stability Results**

The robust stability properties are shown in figure 3.15, the values of $\mu(M_{11})$ in the complex-real uncertainty case are plotted along the frequency axis.

The controller $K_{nom}$ achieves the best robust stability level since there is no loop closure ($K_f = 0$), and apparently there is no perturbation with norm $\leq 1$ that can destabilize the nominal plant.

For $K_2$ robust stability is achieved since $\mu(M_{11}) < 1$. Note that for $K_2$ and $K_3$ two bounds are visible at lower frequencies; they arise from real $\mu$ approximations by optimizing an upper bound and a lower bound. The exact value of $\mu$ lies in between these bounds.

If for $K_2$ only the complex perturbation is taken into account (not shown) the plot moves approximately a level of 0.1 downwards, which means that there is a 10 % stability robustness degradation to unmodelled specifications in the design.

It is remarkable that the robust stability level for $K_3$ is worse than the one of $K_2$. In a small frequency range it is even possible to find a combination of perturbations ($\bar{\sigma}(\Delta_i) \leq 1$) that destabilizes the plant ($\mu(M_{11}) > 1$). This is due to the fact hat the peak value of $\mu(M(j\omega))$ has been optimized. Taking the parametric uncertainties into account in the design improves this value compared to the $\mu$-test for $K_2$. In this sense third design has been successful. The balance between performance and stability robustness has changed. The performance has improved at the cost of stability robustness, while the overall robust performance level has improved.

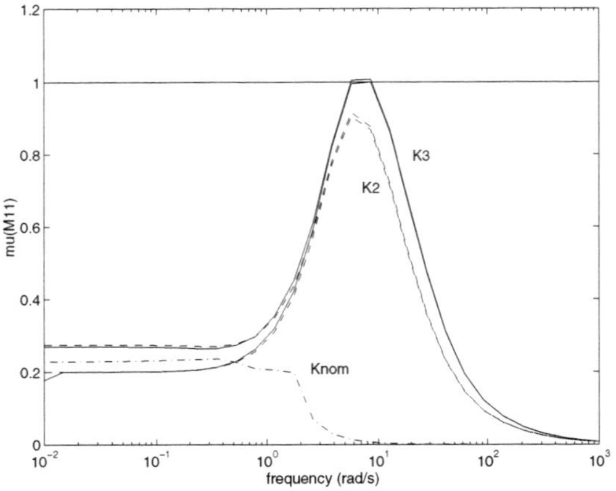


Figure 3.15: Robust stability controllers

## 3.7 Time Simulations: Performance Sensitivity

To establish the practical relevance of the robust performance prediction coming from the $\mu$ anaylsis a time domain analysis is performed on the closed-loop system.

When no perturbations are placed on the system, which is the pure nominal performance case, the pitch rate command responses for all designs match exactly the ideal model response. This means that the frequency domain weights have captured the desired time domain model properties.
The simulation results are shown in figure 3.16.

In order to demonstrate the performance robustness properties three perturbation cases are considered:

- *nom*: simulation with the nominal model;
- *pert1*: simulation with a perturbed model: all parameters in table 3.1 are perturbed with $\delta = -20\%$; only $M_{\delta_e}$ with $\delta = +20\%$;
- *pert2*: identical perturbation, now $\Delta M_{\delta_e} = -20\%$ the others are 20%.

The first plot shows the response of nominal controller $K_{nom}$. When there are no perturbations the nominal system response coincides exactly with the ideal model response. The perturbations have dramatic effect on the tracking performance. The poor robustness property has been predicted by the $\mu$ analysis.

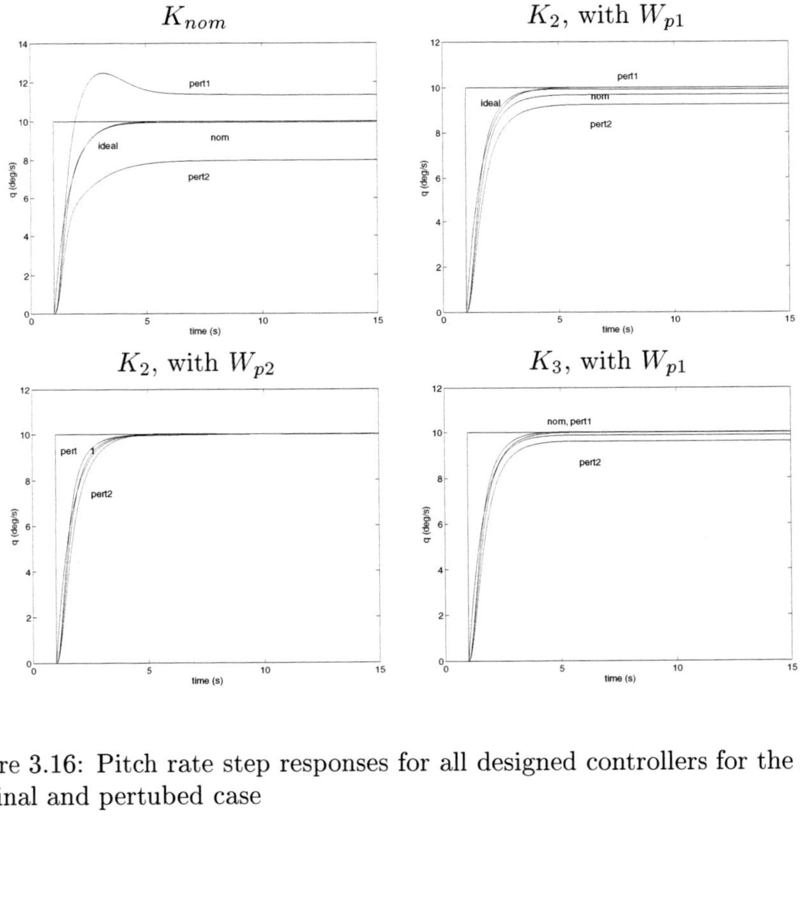


Figure 3.16: Pitch rate step responses for all designed controllers for the ideal, nominal and pertubed case

The simulations results comfirm that a feedforward controller is not dedicated to cope with any model uncertainty or plant disturbance.

The effect of feedback is demonstrated in the responses for the other controllers. The effects of perturbations are less dramatic.
Note that for the second controller $K_2$ designed with $W_{p2}$ the steady state error indeed has become nearly zero, even under the influence of the perturbations. Finally, we can see that $K_3$ performs better than $K_2$ (with $W_{p1}$) under the perturbations, as could be expected from the $\mu$-analysis. Further the controller $K_3$ with $W_{p2}$ not shown here has a similar behavior to the controller $K_2$ designed with $W_{p2}$. Finally, we remark even though not shown here that the control activity requirements have all been met for all designs and perturbation cases. The peak elevator deflections were within a factor 4 and the deflection rates within a factor 3 with respect to the specified bounds.

### Time Simulations, Robust Stability Aspect

Finally, let us determine the exact perturbation combination that could destabilize the closed-loop systems for the controller case $K_2$ and $K_3$. As we see from figure 3.15, the $\mu$ plots (lower bounds) of the closed-loop systems show peak values of $\mu_{peak} = 0.918$ at $\omega_{peak} = 6.5\ rad/s$ and $\mu_{peak} = 1.024$ at $\omega_{peak} = 6.9\ rad/s$ respectively. This means that we can find the smallest destabilizing perturbation with appropriate $\Delta$-*structure* given by:

$$\Delta = \{diag[\delta_1,\ \cdots,\ \delta_6] : \delta_1,\ \cdots,\ \delta_5 \in \mathbb{R},\ \delta_6 \in \mathbb{C}\}$$

with $\bar{\sigma}(\Delta) = 1/0.918 = 1.089$ and $\bar{\sigma}(\Delta) = 1/1.024 = 0.977$ respectively. The system with $K_2$ is robustly stable, since $\mu_{peak} < 1$ and the norm of the perturbation therefore needs to be larger than 1 to destabilize the system. This is not the case for $K_3$. Using available software we have found perturbations that will just destabilize the systems. For $K_2$ such a destabilizing perturbation is given by:

$$\begin{aligned}\Delta &= diag[\delta_1,\ \cdots,\ \delta_6] \\ &= diag[-1.0892,\ -0.8389,\ 0.7893,\ -1.0892,\ 1.0892,\ -1.0573 - 0.2618i]\end{aligned}$$

with norm $\bar{\sigma}(\Delta) = 1.098$. For $K_3$ the perturbation is:

$$\Delta = diag[-0.9768,\ -0.9768,\ 0.0073,\ -0.9768,\ 0.9768,\ -0.9624 - 0.1667i]$$

In figure 3.17 and 3.18 the closed-loop system is simulated without and with perturbations. In order to determine how sensitive the system is with respect to the destabilizing perturbation we implement also the perturbation scaled to 98 % and 1.02 % of its critical value. We can see that both systems with controllers $K_2$ and $K_3$ indeed are faster destabilized while increasing the perturbation levels over their critical values. For perturbations below critical the system remains stable as predicted by $\mu$. The result not shown here is that the controller $K_3$ designed with $W_{p2}$ has a lower robust stability level than when designed with $W_{p1}$. This behavior is expectable since the thighter performance requirement has to be exchanged at the expense of robust stability.

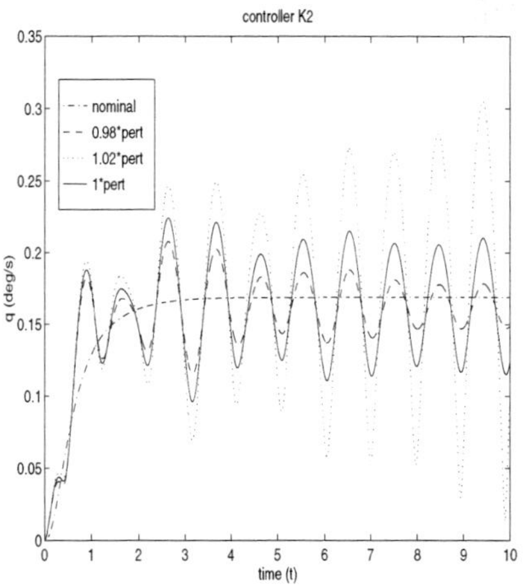


Figure 3.17: Pitch rate step responses nominal, with destabilizing perturbation and ±2% offset perturbation for controller $K_2$

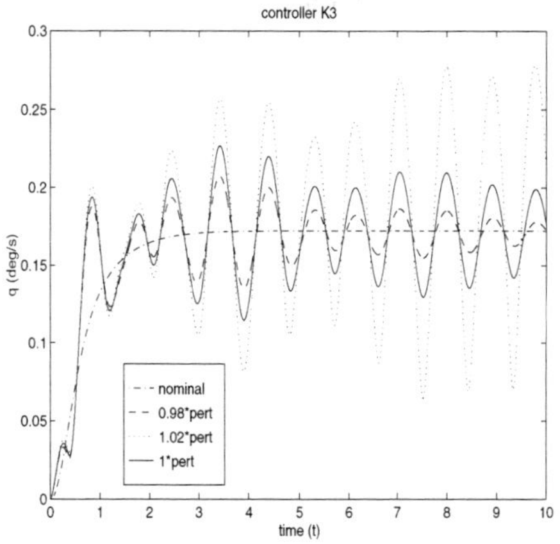


Figure 3.18: Pitch rate step responses nominal, with destabilizing perturbation and ±2% offset perturbation for controller $K_3$

## 3.8 Conclusion

Although we have studied a fairly simple system the example shows that we can deal uniformly with multiple spefications and that the $\mu$ framework is a powerful design tool aimed at obtaining robustly performing solutions. We have seen that the method allows the designer to take into account many factors directly from physical plant considerations. This is because we have the freedom to place the design requirements on physical signals while deliberate uncertainties can be introduced and specialized at several levels in the system components. Finally, as in the experimental world all signals need to be scaled appropriately with respect to each other in order to produce a meaningful design.

Our design approach is to start with the simplest possible design, that is the nominal performance design. In the absence of uncertainty, the two degrees of freedom controller inverts the plant in the feedforward path while leaving the feedback path almost open. This strategy is desirable only in the absence of uncertainties. We have seen that in the nominal design the presence of the slightest uncertainty causes a substantial performance degradation. From a robustness perspective the nominal design is useless. But it has to be conducted in first instance in order to establish the performance drivers in the system. A detailed analysis allows us to shape the initial weighting functions for the nominal performance matrix $M_{22}$.

In a next design step, we gradually add uncertainty into the design. First, we asssume only one general complex uncertainty for various performance levels.

Once this design works we can add the more specialized uncertainties to the system.
Due to the uncertainties the feedback loop becomes increasingly important over the high frequency range. The effect of increasing performance places more control emphasis in the low frequency range. The overall shape of the controller has high gain at low frequencies and rolls-off at higher frequencies which corresponds to the ideas of classical frequency domain loop shaping. This is important because the control solution can be easily interpreted in a industrial environment.

The often reported difficulty in designing a good controller with the right trade-offs between various conflicting requirements has been overcome. This is due to the guidelines provided by the strong $\mu$ analsyis results for the achievement of robust performance. The underlying robust stability and nominal performance results have guided us in the trade-off decision making. To achieve these results we have used $\mu$-synthesis which allows us to alleviate the conservativeness of the $H_\infty$ synthesis method by taking into account stuctured perturbation. The price to be payed for robust performance is an increasing size of the controller state dimension resulting from the $D$-scales. Often their dimension is too high for implementation and one has to resort to model reduction techniques. However we would like that after model reduction the inherited robustness properties remain if we actually need to physically implement the control. This is why we recommend to establish new performance preserving model reduction techniques. The development of such tools has been studied by Goddard and Glover [GG98] and should be integral part within the $\mu$-framework and the commercial available software [BDG$^+$93a]. Such a tool is of crucial importance even in the case we have access to high performance digital signal processors.

Finally, to be successful in improving the behavior of complex systems the method forces the designer to exactly quantify the system specifications while relying continuously on a better and deeper system knowledge. In robust control the design task has shifted from controller design towards specification and component design.

# Chapter 4

# $\mu$-Synthesis: RCAM $\mu$-Design

*This chapter presents a $\mu$-synthesis application to the RCAM autopilot design problem. The autopilot should robustly provide in an automatic approach to land capability along a desired track. It is shown how to translate the RCAM performance and robustness objectives into a $\mu$ optimization problem for the achievement of robust performance.*
*To demonstrate another flexiblity of the method, various multivariable control architectures have been chosen for the control of the longitudinal and lateral aircraft motions. The evaluation of the controllers results in a satisfactory robustness and performance behavior.*

## 4.1 Introduction

The mission and scenario to be 'flown' by the controlled RCAM consists of a landing approach divided into the following segments (see figure 4.1). To be able to evaluate all kinds of different control design procedures and resulting controllers it is necessary to find a uniform evaluation procedure, independent of the design method. An established procedure to do this is to define a mission and a typical landing approach scenario (see [Wea92, Bru94, YJ94]). This mission consists of manoeuvres that can be evaluated by means of nonlinear simulations. The performance of the control law depends on the mission phase, within which hard criteria or bounds on certain signals should be met and/or error signals must be minimized. Satisfactory behavior of the controlled system is evaluated along the prescribed trajectory.

**Segment I (0 to 1).**
Starting at an altitude of 1000 *m* and with a track angle of $\chi = -90$ *deg*, level

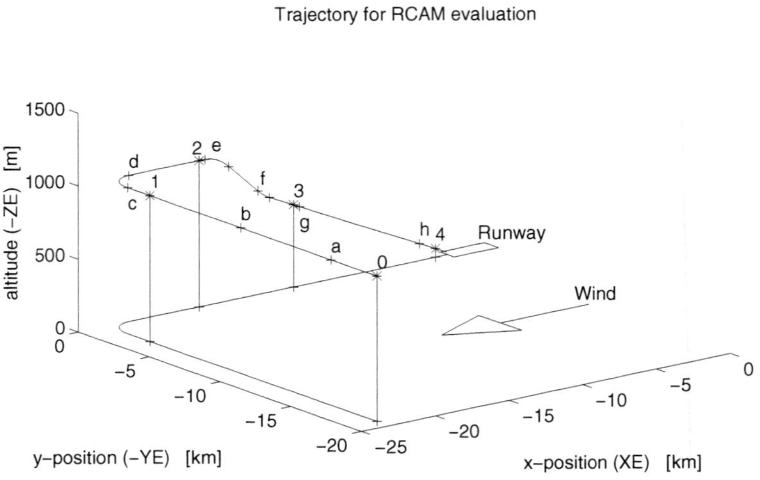


Figure 4.1: The landing approach for the RCAM

flight is to be maintained with a constant airspeed of 80 $m/s$[1]. During this segment, the lateral features of the autopilot will be investigated by simulating failure of the left engine (engine 1). This is indicated in figure 4.1: the failure occurs at point **a**, after which the engine is restarted at point **b**. The transient and steady-state behavior of the system will be analyzed.

**Segment II (1 to 2).**
This segment consists of a commanded co-ordinated turn from points **c** to **d** with a heading rate of $\dot{\psi} = 3$ *deg/sec*. The objectives are to maintain a constant speed of 80 $m/s$, to keep the lateral acceleration close to zero, and to restrict the bank angle to $\phi = 30$ *deg* with consistent rudder/aileron deflections.

**Segment III (2 to 3).**
The descent phase will be started according to the so-called *Frankfurt* descent procedure (see [Bro94]), which has been proposed for reasons of environmental noise reduction. This descent procedure is engaged later and is steeper than the classical descent, which has a constant glide slope angle of $\gamma = -3$ *deg*. The starting altitude is $h = 1000$ $m$. After a short period of level flight, the flightpath angle is set to $\gamma = -6$ *deg* at point **e**, and to $\gamma = -3$ *deg* at point **f**. The desired airspeed is 80 $m/s$.

**Segment IV (3 to 4).**
The glide slope of $\gamma = -3$ *deg* is to be maintained during a wind shear between

[1]The nominal landing airspeed depends on the aircraft mass, it is taken equal to 1.3 times $V_{\text{stall}}$: with a maximum landing mass of 150000 $kg$ this results in $\approx 80$ $m/s$.

points **g** and **h**. The aircraft has to maintain safe flight and should not deviate too far from the desired glide path. The wind shear model used in the evaluation procedure is a two dimensional model derived from [Rob91] (also see subsection A.6.4 for more information). The desired airspeed is 80 $m/s$.

Throughout the mission a Dryden turbulence field, of scale length $L = 305\,m$ and amplitude $\sigma = 0.08$ $m/s$ is acting on the aircraft. Note that the amplitude is only 5% of the amplitude for moderate conditions. This value has been chosen in order to prevent that the effect of turbulence on lateral and longitudinal accelerations overrules other effects of interest. Superimposed on top of the turbulence is a 10 $m/s$ constant wind with a fixed heading. This constant wind is active in full respect during Segments I and II until point **d**, and is slowly reduced to zero between points **d** of Segment II and **g** of Segment IV (at the start of the wind shear model). The wind has no vertical component and is directed along the negative earth-fixed $x$-axis, i.e., it is a cross-wind during Segment I and a headwind during Segment III.

To check robustness properties the entire approach will be flown with a most forward, a nominal and a most aft horizontal centre of gravity location. Furthermore, one flight will be executed with a nominal centre of gravity location and a time delay of 100 $ms$.

The controller design problem for the RCAM model is characterized by a number of fundamental trade-offs between conflicting design specifications. For typical aircraft autopilot systems five classes of criteria are recognized:

- **Performance Criteria**: these reflect tracking error and disturbance rejection characteristics to external signals;
- **Robustness Criteria**: these reflect the stability bounds with respect to parameter variations;
- **Ride Quality Criteria**: these reflect the desire to obtain sufficient passenger and pilot comfort in the form of bounds on certain maximum allowable accelerations and minimum damping levels;
- **Safety Criteria**: these reflect envelope safeguards;
- **Control Activity Criteria**: these are a measure of the power consumed by the controls and also give an indication of fatigue effects.

The theoretical background of $\mu$-analysis and synthesis was discussed in chapter 2. $\mu$-Synthesis is a multivariable closed-loop design methodology addressing the robust performance issue. The autopilot design problem with the above mentioned requirements is reformulated as a $\mu$-optimization problem.
Succesfull application of the method to the RCAM problem needs both, flight mechanical insight and classical Bode shape background. Flight mechanical

insight leads to a sensible problem formulation in the choice of the control architecture not directly provided by control theory. This insight is also needed to understand the problem specifications given appendix A. The various control architectures partly solving our problem are given in section 4.2.

In section 4.3 we show how to reformulate the design requirements into a mathematical format suitable for design and analysis. This is what we call the interconnection structure consisting of the initial feeback architecture of section 4.2 augmented with the requirements. The design specifations are translated mathematically into weighting functions. Weighting functions reflect the frequency responses of the desired closed-loop transfer functions. Within the $\mu$ analysis framework these weighting functions can be constructed and altered with some loop shaping background. The actual controller synthesis given section 4.4. Emphasis in the design is placed on the weighting function selection. The design activity has moved from producing a controller towards producing specifications. This in turn requires the ability for physical interpretation of the results and the requirements.

For the longitudinal motion an energy study is conducted to analyze the interaction between airspeed and altitude wich we would like to control simulatneously through appropriate throttle and the elevator commands. Understanding these dynamics through analysis in the $\mu$-framework has guided us on how to achieve decoupled altitude and speed responses with minimum actuator effort and a pilot like co-ordination of the controls.

For the lateral motion we have studied several trim options for the engine failure case. This study lead us to opt for the aerodynamically best solution wich we called zero sideslip option. In sections 4.5 and 4.6 simulations and interpretation of the results are presented.

## 4.2 Selection of the Control Architecture

The general breakdown of the controller architecture is classical in the sense that separate controllers for the longitudinal and lateral motions have been designed. The two main tasks for both controllers are stability augmentation (innerloop control) and guidance along a flightpath by tracking altitude, lateral position and airspeed references: $h_{ref}$, $y_{ref}$ and $V_{A_{ref}}$ (outerloop control).

### Longitudinal Controller

The basic architecture of the longitudinal controller is depicted in figure 4.2. This system uses the tailplane deflection ($\delta_T$) and the throttle positions ($\delta_{TH1,2}$) to control longitudinal aircraft dynamics (A/C-block). The inputs of the controller $K_{lon}$, consists of attitude angle $\theta$, pitch rate $q$ (providing for innerloop bandwidth and short period damping respectively) and errors in airspeed and altitude and their derivatives. In the requirement the attitude $\theta$ is not available as a measurement. For control this signal is obtained from integration of the measured pitch rate $q$.

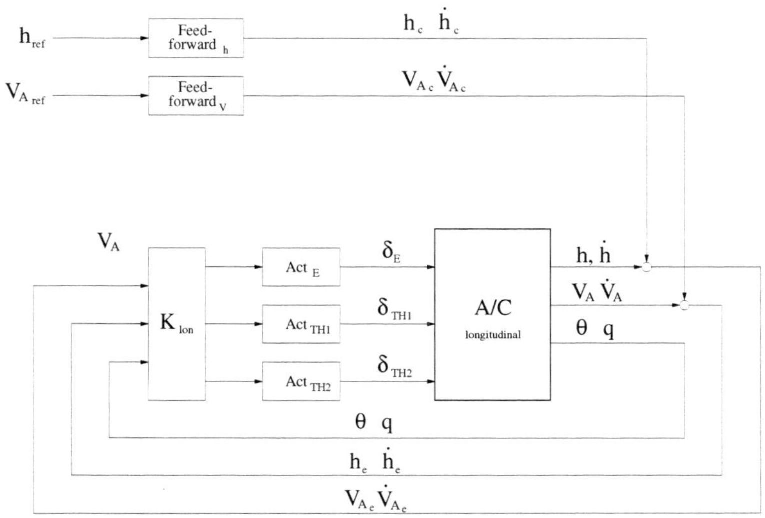


Figure 4.2: Longitudinal controller architecture

The two degrees of freedom control structure consists of a feedback controller and of feedforward filters, generating smooth altitude and speed commands. These feedforward filters are defined by the designer, while the feedback controller is obtained from optimization via DK-iteration, as described in chapter 2 and implemented as a system in state-space form.
The closed-loop system has to track the filtered altitude and speed references as tightly as possible. In fact, the input filters represent ideal models of the closed-loop system.
This architecture allows the input filters to be replaced by some other generator of smooth signals, like the trajectory generator in the evaluation procedure, described in section 4.1.
Note that the feedback controller combines both innerloop as well as guidance functions.

### Lateral Controller

The lateral aircraft dynamics are somewhat more complicated, since the motions have to be considered in both the $x_B - y_B$-plane (e.g. heading) and the $y_B - z_B$-plane (e.g. roll). In the longitudinal case, attitude and position changes only occur in the $x_B - z_B$-plane. To control the lateral motions in a co-ordinated way, a slightly different architecture is applied as shown in figure 4.2.
Separate innerloop and outerloop systems are designed. The innerloop controller will provide stability augmentation and has to track roll and sideslip angle commands. The inputs are roll and sideslip commands and measurements and

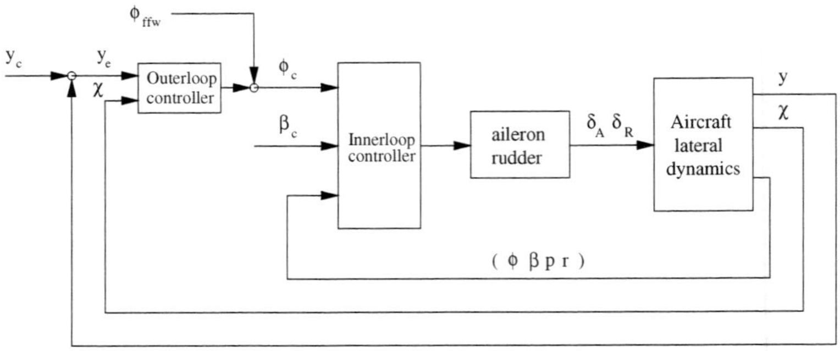


Figure 4.3: Lateral controller architecture

roll and yaw rates, while the outputs are aileron and rudder commands. A sideslip angle could be commanded in case of wing-low cross-wind approach, however during most flight situations, including an engine failure, the sideslip angle will be regulated to zero. Roll angle commands are provided by the outerloop controller, with the task of lateral guidance.

A turn (in zero wind conditions) is characterised by a steady-state roll angle. This architecture allows a roll angle to be commanded in a feed-forward path, improving tracking performance in turns.

Although less explicit than the longitudinal system, the lateral innerloop controller has also a two-degrees of freedom structure. This is achieved by feeding full command and corresponding feedback signals to the controllers, instead of the errors only.

The outerloop controller minimizes lateral position errors via roll commands to the innerloop controller. Feedback of the heading angle improves damping of closed-loop lateral position dynamics.

## 4.3 Translation of the Design Criteria into Method Dependent Objectives

In this section the translation of design criteria into the $\mu$-framework is described. We will concentrate on the design of the longitudinal controller.

### Design Criteria in the General Framework

In the design specifications performance, safety, actuator effort and comfort requirements are distinguished. As we will see, these can be put in a single performance vector. In order to reflect the robustness issue the nominal plant can be extended to a set of plants we design for, using the perturbation channels ($z$, $w$, with $w = \Delta z$).

In figure 4.4 we see that all performance variables can be observed by looking

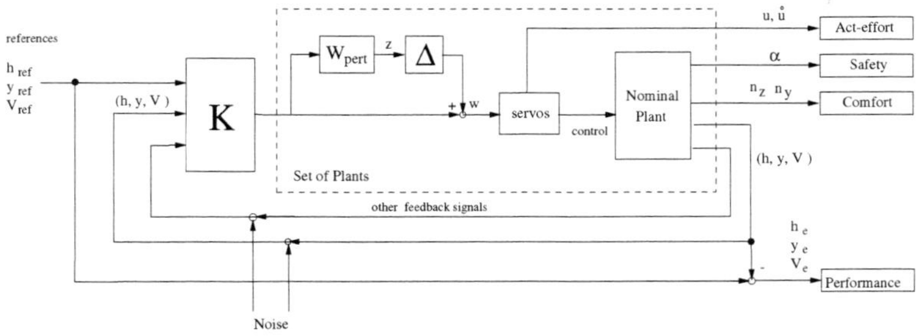


Figure 4.4: Translation design requirements I

at *the signals of interest* at their *physical locations* in the loop. For example, to watch actuator effort, we can pull out the control signals at the servos. For safety, we would like to look at the angle-of-attack.

As a basis for $\mu$-analysis and synthesis, we can in this representation simultaneously take care of the specified design objectives.
The RCAM problem formulation implemented in the $\mu$-design framework can be divided into the following subproblems:

1. **Robust Stability (RS):**
   The closed-loop system should be stabilized for the specified set of plant models. For example (input multiplicative uncertainty):

   $$\tilde{G}(s) = \{G_0(s)\,(1 + W_{pert}(s)\Delta(s)) \mid \|\Delta\|_\infty \leq 1\}$$

   where the operator $\Delta$ is norm bounded and where $W_{pert}(s)$ is a diagonal transfer function matrix shaping the magnitude of the uncertainty at each actuator input along the frequency axis.

   The transfer function $G_0(s)$ represents the linearized aircraft dynamics in the nominal design condition.

2. **Nominal Performance (NP):**
   The closed-loop system should satisfy the given tracking performance, safety, actuator effort and comfort requirements for the nominal plant $G_0(s)$.

3. **Robust Performance (RP):**
   The closed-loop system should achieve the required performance objectives over the whole set of plants defined by $\tilde{G}(s)$.

Next, we transform the representation in figure 4.4 to the framework depicted in fig. 4.5. We pull the *uncertain element* $\Delta$ out of the system, as well as the *sub-system to be designed*, the controller $K$. To this end we define two new sets of inputs and outputs to the general open-loop plant $P$.

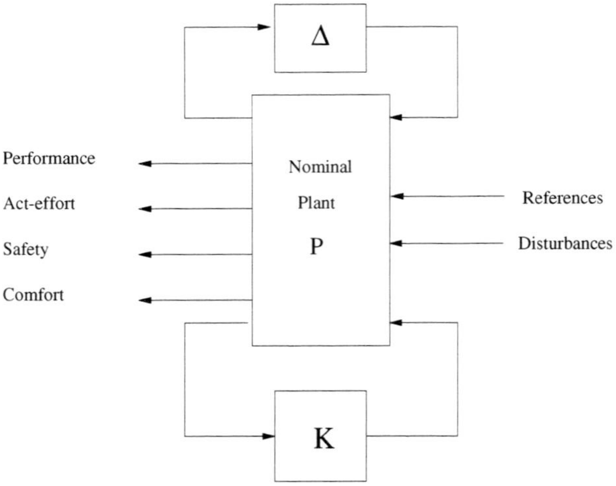


Figure 4.5: Translation design requirements II

Thusfar, we have indicated the closed-loop quantities of interest in a qualitative way. The next step is the quantification of the design requirements using approriate weighting functions for each signal we wish to penalize.
First, all input signals are normalized with respect to their maximum value. In other words, this means that the input signals have magnitude 1 and need to be scaled to a suitable magnitude (relative to other inputs) before entering the system. In this way physical system conditioning is provided.

At the outputs of the performance channel we add constant or frequency dependent weighting functions specifying the relative importance of keeping the signal of interest small. In a same fashion we add filters to the inputs of $\Delta$ to shape the magnitudes of the perturbations.
The *weighted* closed-loop system now looks like:

$$M = W_{out} F_l(P, K) W_{in}$$

where $W_{in}$ and $W_{out}$ contain input and output weighting functions on the signals, including scalings, performance weights, perturbation filters, noise-shaping filters reference filters etc. This is shown in figure 4.6.
Finally we can apply the general theorems for analysis:

1. **Robust Stability** $\mu(M_{11}(j\omega)) < 1 \; \forall \, \omega$
2. **Nominal Performance** $\bar{\sigma}(M_{22}(j\omega)) < \; \forall \, \omega$
3. **Robust Performance** $\mu(M(j\omega)) < 1 \; \forall \, \omega$

In figure 4.6 the groups of signals involved in each of the theorems have been indicated (RS, NP, RP).

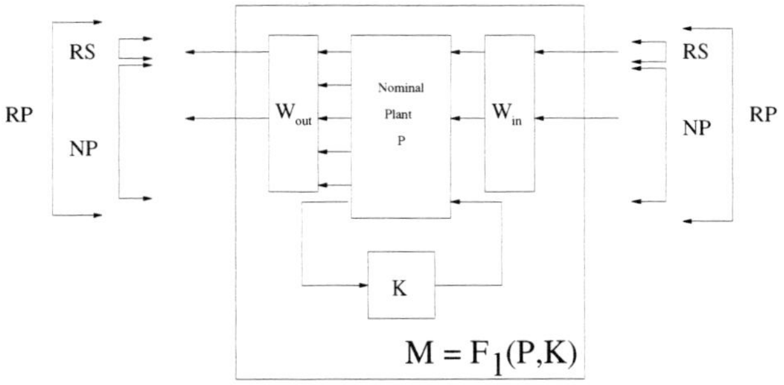


Figure 4.6: Translation of design requirements III

**Application to the Longitudinal Controller**
For the longitudinal controller the main performance objective is to achieve decoupled tracking of speed and altitude commands ($V_{ref}$, $h_{ref}$), see figure 4.7.

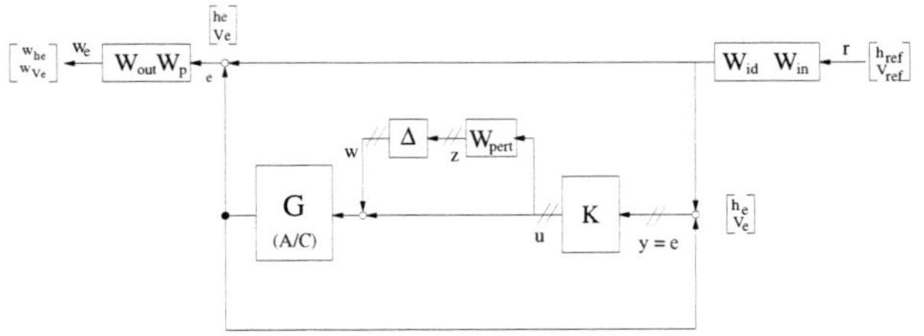


Figure 4.7: Simplified interconnection structure

The commands that enter the controller are smoothed by an ideal second order model, given by input filter matrix, $W_{id}$. We want the closed-loop system to track the filtered references as close as possible. How good this should be done is reflected by the weighting filter matrix $W_p$ placed on the error signals between ideal model response and the actual altitude and speed responses.

In order to normalize the reference and output signals and to shape the interaction level between the altitude and speed channels, we use input and output scaling matrices $W_{in}$ and $W_{out}$, usually taken as $W_{out} = W_{in}^{-1}$.

To simplify the initial design and the analysis of the problem we have captured the model uncertainties such as mass, centre of gravity variations and computational time delay by an input multiplicative uncertainty model with weighting function $W_{pert}$.

The first step in the analysis is to derive from figure 4.7 the general open-loop interconnection structure $P$ as given in figure 4.6.

1. **Open-loop Interconnection Structure $P$**

In figure 4.7 the uncertainty $\Delta$ is an internal variable. We want to have the uncertainty affecting $P$ in a feedback way as shown in figure 4.6.

Therefore introduce at the input and output of $\Delta$ the signals $z$ and $w$ respectively. Next, pull the $\Delta$ out of the closed-loop system as shown in figure 4.8.

In the same way, we can pull out the controller $K$ by defining $y$ and $u$ as the controller input and output respectively and cut the lines at these spots.

The remaining channels are the performance channels of $P$ with input signals $r = [h_{ref}\ V_{ref}]^T$ and output signals $w_e = [w_{h_e}\ w_{V_e}]^T$.

When disconnecting the controller and the uncertainity from the closed-loop structure we can rewrite the equations for the partitioned open-loop system $P$ from figure 4.8 as:

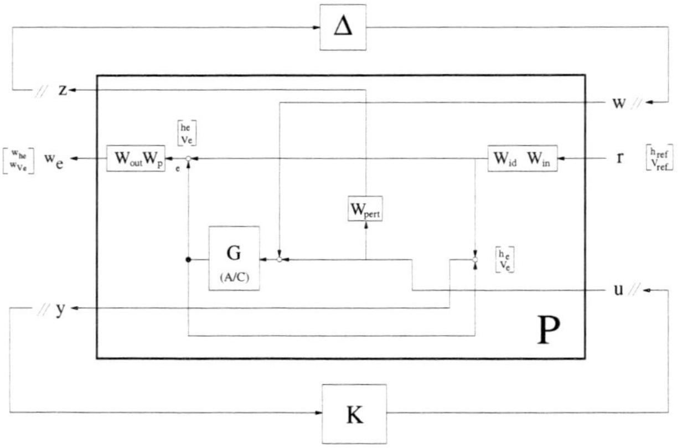


Figure 4.8: Transformation to the general system representation

$$\begin{bmatrix} z \\ w_e \\ \hline y \end{bmatrix} = \left[\begin{array}{c|c} P_{11} & P_{12} \\ \hline P_{21} & P_{22} \end{array}\right] \begin{bmatrix} w \\ r \\ \hline u \end{bmatrix}$$

$$= \left[\begin{array}{cc|c} 0 & 0 & W_{pert} \\ -W_{out}W_pG & W_{out}W_pW_{id}W_{in} & -W_{out}W_pG \\ \hline -G & W_{id}W_{in} & -G \end{array}\right] \begin{bmatrix} w \\ r \\ \hline u \end{bmatrix} \tag{4.1}$$

2. **Closed-loop Interconnection Structure** $M$

To obtain the closed-loop structure $M = F_l(P, K)$ we apply the definition of the lower linear fractional transformation:

$$M = F_l(P, K) = P_{11} + P_{12}K(I - P_{22}K)^{-1}P_{21}$$

Using the definitions: $S_o = (I+GK)^{-1}$, $T_i = KG(I+KG)^{-1}$ [Mac89] and the relation, $K(I + GK)^{-1}G = (I + KG)^{-1}KG$ ("push-through rule"), we find the desired relation for analysis:

$$M = \begin{bmatrix} -W_{pert}T_i & W_{pert}KS_oW_{id}W_{in} \\ W_{out}W_pS_oG & W_{out}W_pS_oW_{id}W_{in} \end{bmatrix}$$

3. **Robust Stability**

As can be seen from the matrix $M$, robust stability is characterized by a weighted input complementary sensitivity $M_{11} = W_{pert}T_i$. Robust stability is achieved whenever the following condition is satisfied:

$$\mu(W_{pert}T_i) < 1.$$

4. **Nominal Performance**

The performance channel is characterized by a weighted output sensitivity $M_{22} = W_{out}W_pS_oW_{id}W_{in}$. Let us concentrate on the specification of decoupled tracking of altitude and speed commands by analyzing the performance block $M_{22}$ in more detail:

$$\begin{bmatrix} w_{e_h} \\ w_{e_V} \end{bmatrix} = \begin{bmatrix} W_{out_h}W_{p_h}\frac{h_e}{h}W_{id_h}W_{in_h} & W_{out_h}W_{p_h}\frac{h_e}{V}W_{id_V}W_{in_V} \\ W_{out_V}W_{p_V}\frac{V_e}{h}W_{id_h}W_{in_h} & W_{out_V}W_{p_V}\frac{V_e}{V}W_{id_V}W_{in_V} \end{bmatrix} \begin{bmatrix} h_{ref} \\ V_{ref} \end{bmatrix}$$

Usually we take $W_{out} = W_{in}^{-1}$:

$$\begin{bmatrix} w_{e_h} \\ w_{e_V} \end{bmatrix} = \begin{bmatrix} W_{p_h}\frac{h_e}{h}W_{id_h} & W_{out_h}W_{p_h}\frac{h_e}{V}W_{id_V}W_{in_V} \\ W_{out_V}W_{p_V}\frac{V_e}{h}W_{id_h}W_{in_h} & W_{p_V}\frac{V_e}{V}W_{id_V} \end{bmatrix} \begin{bmatrix} h_{ref} \\ V_{ref} \end{bmatrix}$$

Taking this choice for the input and output scalings only affects the off-diagonal terms in $M_{22}$. Choosing the input channel scalings as:

$$diag(W_{in_h},\ W_{in_V}) = diag(1/10,\ 10)$$

and the output scalings as

$$diag(W_{out_h},\ W_{out_V}) = diag(10,\ 1/10)$$

does not affect the diagonal terms in the matrix $M_{22}$. The scalings affect only the off-diagonal terms. One off-diagonal term is amplified while the other is reduced with a factor 100. The shown mechanism influencing the interaction between the speed and altitude channels is the same offset mechanism as the $D$-scales have on the performance robustness interaction channels $M_{12}$ and $M_{21}$.

The nominal decoupling tracking objective is achieved whenever the following condition is satisfied:

$$||W_{out}W_pS_oW_{id}W_{in}||_\infty < 1.$$

5. **Robust Performance**

Finally, to obtain the robust performance condition for this problem we require $\mu(M) < 1$. This results in the following $\mu$-upper bound condition:

$$\mu(M) \leq \inf_D \bar{\sigma}\left(\begin{bmatrix} -W_{pert}T_i & DW_{pert}KS_oW_{id}W_{in} \\ W_{out}W_pS_oGD^{-1} & W_{out}W_pS_oW_{id}W_{in} \end{bmatrix}\right)$$

where $D$ are the optimizing $D$-scales.

An upperbound for the robust performance level $\mu(M)$ as given by Zhou [ZDG96] is:

$$\mu(M) \leq \sqrt{\kappa(W_{in}^{-1}W_{id}^{-1}G)}(||W_{pert}T_i||_\infty + ||W_{out}W_pS_oW_{id}W_{in}||_\infty)$$

If $W_{in}W_{id} = I$, the achievable robust performance level is proportional to the square root of the condition number of the plant ($\kappa(G)$). By properly choosing $W_{in}$, we can improve the robust performance level and provide for physical system conditioning.

## 4.4 Control Synthesis Design Cycle

### 4.4.1 Introduction

This section describes the design of the flight controller for the RCAM.
In section 4.2 the architecture of the controller was described. According to this architecture three control sub-systems are designed using $\mu$-synthesis:

- a longitudinal controller for altitude and speed reference tracking, with integrated innerloop and outerloop functions;
- a lateral innerloop controller, for tracking roll and sideslip commands;
- a lateral outerloop controller, for lateral guidance.

The design work for these controllers now consists of the following steps:

- choose aircraft configuration and flight conditions for the design point and obtain a linear model for that point;
- define a *lay-out* for the general interconnection structure, consisting of the plant model, controller, weighting functions, modeled uncertainties;
- iteratively synthesize and analyze controllers and adjust weighting functions. Analysis is based on $\mu$-plots, controller properties (e.g. frequency responses, order) and closed-loop time responses.
- preparation of the controller for implementation: removal of fast poles, balancing and order reduction;
- implementation with the nonlinear aircraft model to perform time simulations.

All these steps together are performed iteratively until the process has converged to a satisfactory solution.

We shall globally discuss the design of longitudinal controller. For a description of the overall design, the reader is refered to the Garteur report [BL96].

All the design work is carried out in MATLAB. For controller design the $\mu$-Analysis and Synthesis Toolbox ($\mu$-Tools [BDG$^+$93b]) is applied. For closed-loop simulations Simulink is used.

### 4.4.2 Selection of the Design Model

Before linearizing the system dynamics for controller design, we can get some insight about the available control power from the trimcurves.
These reflect the control power required for horizontal equilibrium flight as a function of the airspeed.
The throttle and tailplane positions are shown in figure 4.9 for different aircraft masses and center of gravity locations respectively. A complete in depth analysis of the RCAM flight performance characteristics are reported in [LB96].
Figure 4.9(a) reveals that the required tailplane deflection becomes quite large at lower airspeeds. Static longitudinal stability as described in Brockhaus [Bro94] is guaranteed, since the sign of $d\delta_T/dV_A$ is always positive.

More interesting are the throttle trim curves, or *performance diagrams*. These plots show a minimum occuring at the *minimum drag speed* depending on the aircraft configuration and mass.
Flying the aircraft at a lower speed than the minimum drag speed, means it is *flying on the backside of the powercurve*, where a lower trim speed requires a higher throttle setting for equilibrium flight.
For an aircraft equipped with separate autopilot and autothrottle systems, flying on the backside of the power curve results in so called *speed instability*, when only one of these systems is engaged. This is one motivation beside the

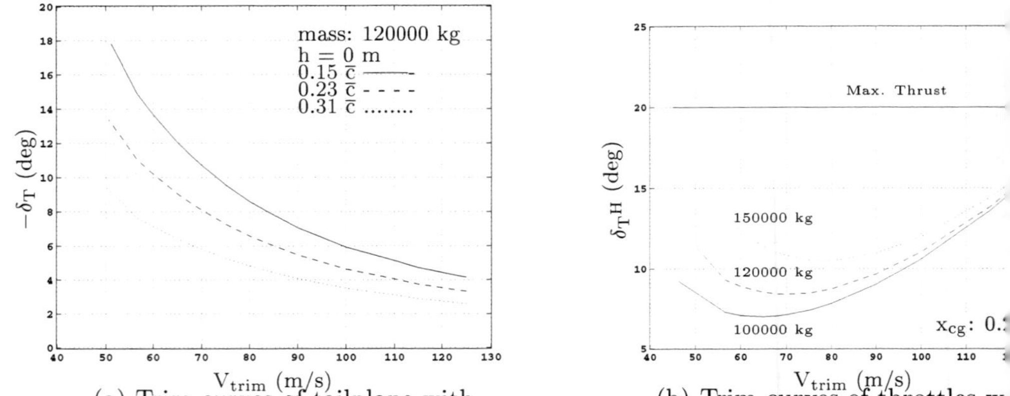


(a) Trim curves of tailplane with different CoG locations

(b) Trim curves of throttles w different masses

Figure 4.9: RCAM trim curves

strong altitude speed interaction for the multivariable problem approach using $\mu$-synthesis which must lead to an integrated autopilot/autothrottle system design.

The nominal design model is obtained by trimming and linearizing the RCAM in a horizontal flight at a speed of 80 m/s, a mass of 120 000 kg, a vertical CoG position of 0 $\bar{c}$ and a horizontal CoG location of 0.23 $\bar{c}$.

An impression of the linear model characteristics can obtained by looking at damping, eigenfrequencies and time constants as shown in table 4.1. Clearly, all modes are stable. The Dutch roll and the short period modes show reasonable damping while the phugoid is badly damped.

| mode | frequency (rad/s) | damping | time constant |
|---|---|---|---|
| phugoid | 0.1269 | 0.0898 | - |
| short period | 1.3837 | 0.6000 | - |
| Dutch roll | 0.6404 | 0.3684 | - |
| Spiral | - | - | 5.4449 |
| aper. roll | - | - | 0.7682 |

Table 4.1: Characteristics of linear design model

### 4.4.3 The Longitudinal Controller

The first step in a $\mu$-synthesis design is to develop an interconnection structure for design. This structure includes the design requirements on top of the chosen basic control architecture. The structure for the longitudinal controller is shown in figure 4.10. Each of the blocks in the diagram will be discussed next.

**The Aircraft Dynamics: A/C**

This block contains the linearized longitudinal aircraft dynamics denoted by $A/C$.

As shown by Lambregts [Lam83c] $\dot{V}_A$ can not be measured. For practical implementation of this signal we synthesize $\dot{V}_{A_{syn}}$ from the airspeed $V_A$ through complementary filtering by combining the total inertial airspeed $V$ with airdata measurements according to:

$$\begin{aligned} \dot{V}_{A_{syn}} &= \tfrac{\tau s}{\tau s+1}\dot{V} + \tfrac{s}{\tau s+1}(V_A - V_{A_{trim}}) \\ V_{A_{syn}} &= \tfrac{\tau s}{\tau s+1}\dot{V} + \tfrac{1}{\tau s+1}(V_A - V_{A_{trim}}) + \tfrac{\tau s}{\tau s+1}(V - V_{trim}) \end{aligned} \tag{4.2}$$

the time constant $\tau$ is further set to 5 and can be adjusted to vary the control activity in turbulence.

**Tailplane and Engine Dynamics: Act./Engine**

The tailplane actuator and engine dynamics are included in the design. These are modeled as first order filters given by:

$$G_{\delta_T}(s) = \tfrac{1}{0.15s+1} \quad \text{and} \quad G_{\delta_{TH12}}(s) = \tfrac{1}{1.5s+1} \tag{4.3}$$

**The Controller $K$**

The controller receives pitch attitude $\theta$ integrated from pitch rate and the pitch rate measurements for control augmentation of the short period dynamics. For regulation the controller uses the altitude and speed errors and their derivatives respectively, $h_e$, $V_e$ $\dot{h}_e$, $\dot{V}_e$. The controller delivers command signals $\delta_{E_c}$ to the tailplane and to the two engine throttles $\delta_{T_{12c}}$.

**The Uncertainty Model: $W_{pert}$**

The diagonal weighting function matrix $W_{pert}$ reflects the uncertainty in the tailplane and the two engine channels. Even though crude this simple model representation has been chosen to account for uncertainties in the aircraft model, without addressing a specific uncertain parameter. With the loop closed via the complex diagonal matrix $\Delta$ normalized to $||\Delta||_\infty \leq 1$. The resulting uncertain dynamics $(I_{3\times3} + \Delta W_{pert})$ define a set of models for which we want to guarantee stability and to achieve the performance specifications.

The set of models has been chosen large enough to account for several uncertainties in the model, like uncertain actuator responses, uncertain aerodynamic parameters, time delays, center of gravity shifts and mass variations.

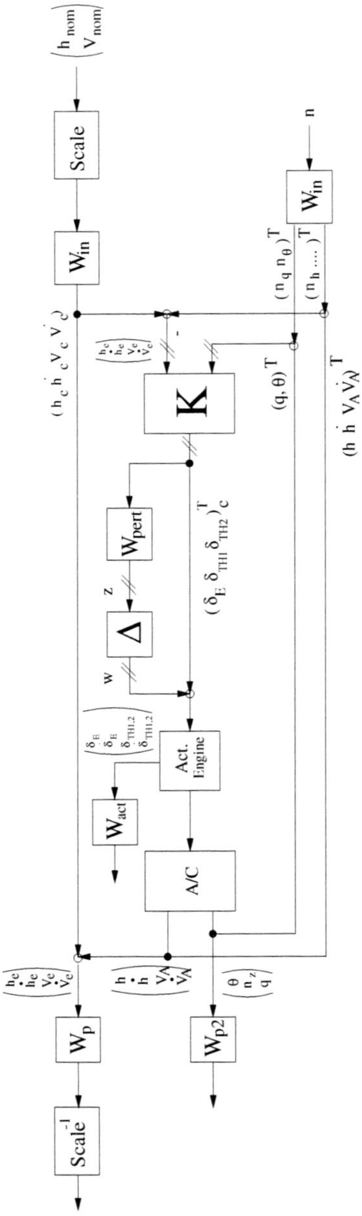


Figure 4.10: Interconnection structure for the longitudinal controller design

Of course the $\mu$-synthesis methodology enables us to address each of these perturbations separately as real parametric variations. This would be appropriate in a later design stage whenever the performance robustness specifications need to be tightened.

At the cost of more design effort in accurately modelling these uncertainities performance will certainly improve, since the parametric model set is less conservative compared to the set adopted for this design.

The weighting functions on the diagonal of the matrix $W_{pert}$ have a low value in the lower frequency range, while in the mid-frequency range the magnitude increases to flatten out at high frequencies. Figure 4.11 shows the frequency responses of the elements in $W_{pert}$. These weightings have been determined after a few iterations.
The shape of this wieghting function forces the controller to roll off at higher frequencies. This is due to the robust stability requirement for which $||T_i||_\infty \leq ||W_{pert}||_\infty$ must be achieved at all frequencies.

$W_{pert}$ is a diagonal matrix: $W_{pert}(s) = \text{diag}(W_{pert_T}(s), W_{pert_{TH1,2}}(s))$ with:

$$W_{pert_T}(s) = 0.25\frac{s+1}{s/200+1} \quad W_{pert_{TH1,2}}(s) = 0.25\frac{0.5s+1}{s/200+1}$$

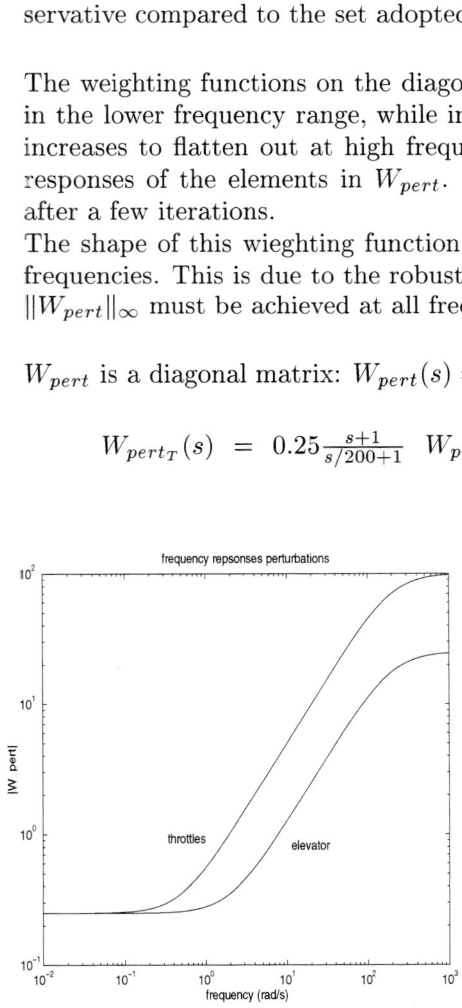


Figure 4.11: Frequency responses of $W_{pert}$

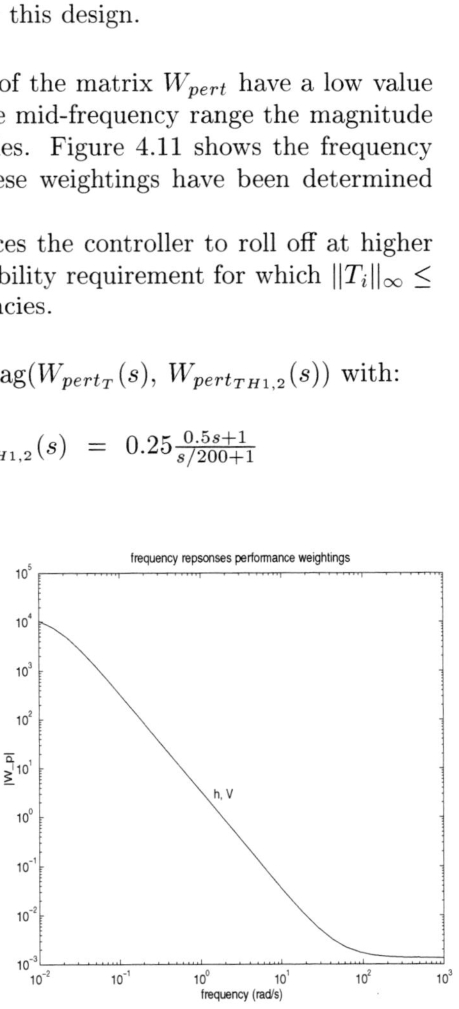


Figure 4.12: Frequency responses of $W_p$

### The Ideal Model Dynamics: $W_{id}$

The block $W_{id}$ contains the feedforward filters that deliver altitude and speed command profiles to references. These are defined by the second order dynamics:

$$\frac{h_c(s)}{h_{ref}(s)} = \frac{V_c(s)}{V_{ref}(s)} = \frac{0.15^2}{s^2 + 2 * 1 * 0.15s + 0.15^2} \tag{4.4}$$

These filters adopted by Kaminer [KBCE90] in previous autopilot Boeing design studies are representative for the desired closed-loop dynamics.

From point-mass dynamics in the gravity field, we know that speed and altitude are correlated via the total energy state of the point-mass (considered constant). To achieve the lowest possible throttle activity, it is necessary to choose identical bandwidths for both filters. More details on the choice of the filter bandwidth and control activity will be given in section 4.5.

### Tracking Error Requirement: $W_p$

Beyond the bandwidth of the ideal model filters, the differences between the altitude and speed responses and respectively the commanded $h_c$ and $V_c$ should be small. This requirement is reflected by the diagonal weighting matrix $W_p$. At lower frequencies the magnitude of $W_p$ is small, while it rolls off at higher frequencies. For both the altitude and speed channels, the weighting functions are taken as:

$$W_{p_h} = W_{p_V} = 15000 \left( \frac{1/50s + 1}{1/0.015s + 1} \right)^2 . \tag{4.5}$$

The weights of the derivatives of the altitude and speed are set to zero. The resulting tracking error weight is given by $W_p(s) = \text{diag}(W_{p_h}(s),\, W_{p_V}(s))$.

To achieve small steady-state errors the system loop gain has to be large at the lower frequencies. This can be achieved by increasing the low frequency gain of the these weighting functions. By moving the pole to the origin and increasing the gain, the cross-over frequency is held constant. This effect has been discussed in detail in [BL96]. In figure 4.12 the frequency responses of $W_p$ are given. Note that weighting contains approximate double integration. From section **??** we have seen that the controller will at lower frequencies have approximately the same shape. Double integration has been included to provide the controlled system with asymptotic tracking properties to ramp signals.

### The Scaling Filters: $W_{in}$ and $W_{out}$

We motivated in section 4.3 that we have to scale the the relative importance between the signals to a realistic level. The output scaling on the errors is usually chosen as the inverse of the input scaling, $W_{out} = W_{in}^{-1}$ to normalize back the outputs. The altitude command is scaled to $h_{cmax} = 5\,m$ and the speed command to $V_{cmax} = 1\,m/sec$, $W_{in} = \text{diag}(h_{cmax},\, V_{cmax})$. These scalings provide the designer a mean to handle couplings between the speed and altitude channels while leaving the diagonal channels unchanged.

**Comfort Requirements:** $W_{p2}$

The weighting function matrix $W_{p2} = \text{diag}(W_{p_q}, W_{p_{nz}})$ is applied in order to penalize excesses in the pitch rate $q$ and normal acceleration $n_z$. Good tracking of the feedforward filter outputs should not be at the cost of extreme pitch rates due to bach short period damping. The weight on the pitch rate signal is chosen constant over all frequencies and is set to $W_{p_q} = 57.3/0.5$ since the pitch rate should never exceed 0.5 $deg/s$ for the command levels adopted in the interconnection structure.
The normal acceleration $n_z$ is weighted with $W_{p_{nz}} = 1/0.04$ to keep the acceleration level to commands below 0.04 $g$.

**Actuator Effort Minimization:** $W_{act}$

This weighting function is diagonal and puts constant weights on the tailplaine and throttle controls and their derivatives. The matrix $W_{act}$ is set to:

$$\begin{aligned} W_{act} &= \text{diag}\left(\frac{1}{\delta_T(max)}, \frac{1}{\dot{\delta}_T(max)}, \frac{1}{\delta_{TH1,2}(max)}, \frac{1}{\dot{\delta}_{TH1,2}(max)}\right) \\ &= \text{diag}\left(\frac{57.3}{10}, \frac{57.3}{15}, \frac{57.3}{2}, \frac{57.3}{1}, \right) \; rad^{-1} \end{aligned}$$

The maximum deflections and rates bounded by these weights are within the limits given specified in section **??** of [Fli96].

**Sensor Noise:** $W_n$

Since in practice measurements are always corrupted with noise the diagonal weighting filter $W_n(s)$ reflects the relative importance of the noise along the frequency in the measurement signals. The individual weights are given by:

$$\begin{aligned} W_{n_h} &= 5 * 10^{-4} \tfrac{1/0.01s+1}{1/500s+1} & W_{n_{\dot{h}}} &= 0.02 \\ W_{n_v} &= 2.5 * 10^{-4} \tfrac{1/0.01s+1}{1/500s+1} & W_{n_{\dot{v}}} &= 0.02 \\ W_{n_\theta} &= 0.05/57.3 & W_{n_q} &= 0.025/57.3 \end{aligned} \tag{4.6}$$

### 4.4.4 Controller Synthesis

Based on the interconnection structures defined above, controllers are obtained by optimizing the structured singlular value upper bound using the DK-iteration scheme. Standard routines in $\mu$-Tools [BDG$^+$93b] have been used. The resulting controllers have high orders so that order reduction is nessesary. The controllers also contain very fast poles which are removed before running simulations. The controllers are always checked for sufficient roll-off to prevent the control system from commanding high-frequency signals to the actuators. Finally, the possible occurence of unstable poles in the controller is always checked before applying order reduction via balanced Hankel-singular value truncation.

The controller order after reduction is still very high. For the longitudinal system 24, the lateral innerloop controller 24 and for the lateral outerloop system 8. Much lower orders can be achieved with better reduction methods. These were not at our disposition and we had to rely on the standard tools offered by $\mu$-Tools [BDG+93b].

The $\mu$-plots are the most important indicators in the design process. They give information about which weightings need to be adjusted for design refinments. For the three final controllers $\mu$-plots are in figure 4.13. The peak value of $\mu$ reflecting our robust performance objective along the frequency axis has been minimized via DK-iteration. For the longitudinal and lateral innerloop controller robust performance is not achieved since $\mu_{peak} > 1$. In all cases stability is guaranteed for the whole adopted set of plants since $\mu_{peak}(M_{11}) < 1$. The desired nominal performance objective is not achieved in the longitudinal case. Nevertheless the tracking objective has been met in all cases which was the main performance driver in our problems. In most design cases it was the acceleration requirement that caused performance deterioration.

## 4.5 Analysis of the Resulting Controllers

### 4.5.1 Longitudinal Controller

For the longitudinal system we concentrate on tracking performance and decoupling of the altitude and speed channels. Figure 4.14 shows linear time responses to altitude and speed step commands. The ideal model responses have also been included in the plots. It is clear that the responses and the decoupling are very satisfactory.
An interesting way to evaluate speed-flightpath decoupling and throttle activity is to look at energy principles. Considering the aircraft as a point-mass, we can write the total energy state in zero wind conditions as follows:

$$E_{tot} = E_{pot} + E_{kin} = mgh + \frac{1}{2}mV^2 \tag{4.7}$$

Taking the time derivative, dividing by $mgV$ and with $\dot{h} = V \sin\gamma$, we get the specific energy rate:

$$\frac{\dot{E}_s}{V} = \sin\gamma + \frac{\dot{V}}{g} \tag{4.8}$$

From equilibrium along the flightpath we can derive:

$$\frac{\dot{E}_s}{V} = \frac{\dot{V}}{g} + \sin\gamma = \frac{T - D}{W} \tag{4.9}$$

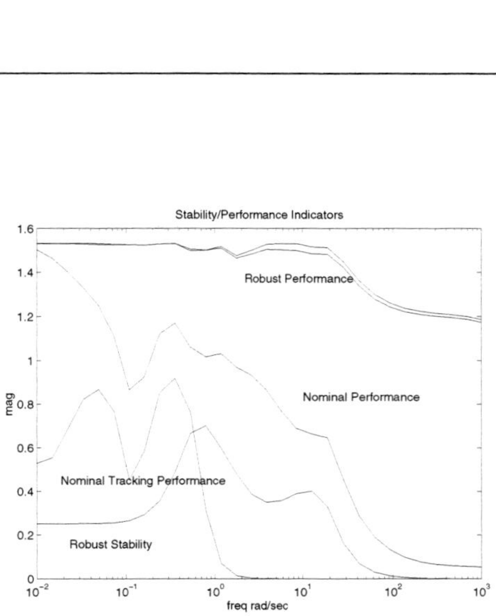


(a) Longitundinal controller

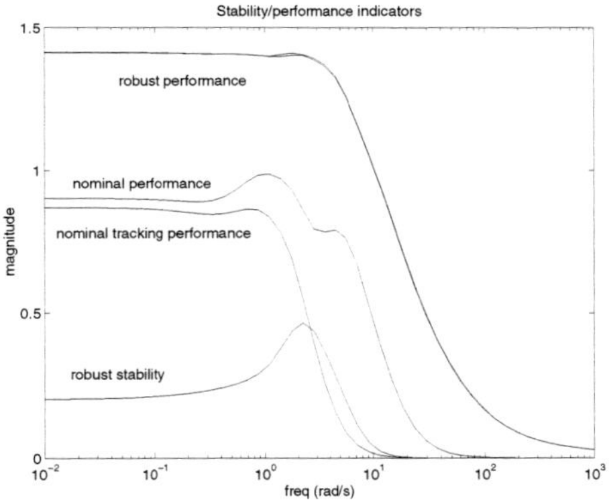


(b) Lateral innerloop controller

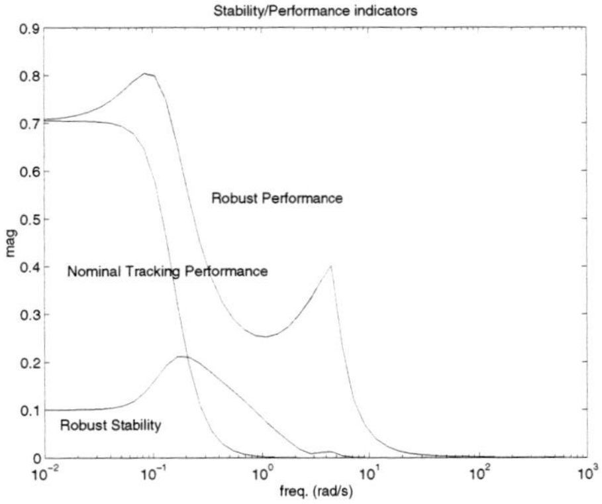


(c) Lateral outerloop controller

Figure 4.13: $\mu$-plots for controllers

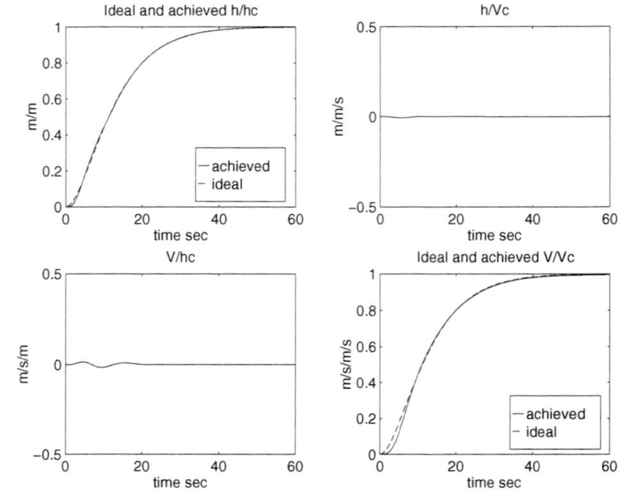


Figure 4.14: Ideal and achieved altitude and speed command responses

Since aircraft preferably fly at their minimum drag speed, the specific energy rate can be controlled by changing the thrust setting.
An interesting way to see whether the controller responds according to this principle, we can define an energy exchange manoeuvre. We command altitude and speed steps so that potential and kinetic energy repectively decrease and increase with the same amount (or vice versa). Since we have chosen identical feedforward filters for both channels (see section 4.2), $\dot{V}/g \approx -\sin\gamma$ during the manoeuvre. This means that $\dot{E}_s \approx 0$ and therefore in the ideal case the throttles should not respond. Since the aircraft is not a point-mass, we expect some response, but this still should be low. The energy exchange is controlled with the (energy conservative) tailplane.

The principle of using throttles for energy input and elevator for energy distribution is the basis for the "Total Energy Control System" (TECS), as described in [Lam83c].

We now command a $-20$ m altitude change and a corresponding 2.42 m/s increase speed (so that $\Delta E_{kin} = -\Delta E_{pot}$, starting $V_A = 80$ m/s).

In figure 4.15 the nonlinear simulation results are depicted. For comparison also the throttle response of an "ad-hoc" controller is given where we required speed/flightpath-angle decoupling adding an extra outerloop to control altitude.

In the third subfigure can be seen that the throttle response is indeed negligible compared to the "ad-hoc" controller. In the fourth figure the specific energy rate and its contributions are depicted. The energy rate remains very small,

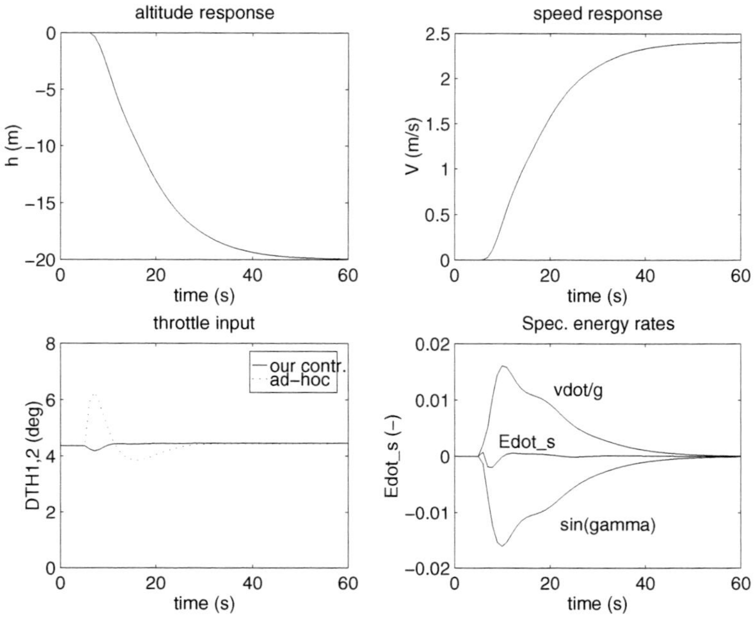


Figure 4.15: Energy manoeuvre reponses

while indeed the $\dot{V}/g \approx -\sin\gamma$.

### 4.5.2 Lateral Controller

Figure 4.16 shows linear time reponses of the aircraft at design conditions with the lateral innerloop controller.
The closed-loop system responses match those of the ideal models very well. Also the achieved decoupling is good. Note that the roll angle shows more response to a sideslip command than vice versa. This can be modified, by tuning $W_{in}$ and $W_{out}$, but this not is considered necessary. Figure 4.17 shows time reponses at the design conditions for the linear model and the ideal model. The closed-loop response has a considerable time lag compared to the ideal model. However, we also have to look at the control activity. For safety and comfort reasons, the innerloop controller should be able to follow the commands from the outerloop controller smoothly, see figure 4.18. In other words: the commands from the outerloop controller should allow the aircraft to "roll in" for a heading change. Therefore we have to accept this time lag.
For a more detailed analysis of the control system, we refer to the design report, [BL96]. In the following section the results of the automated evaluation procedure (see section 4.1) are discussed. [BL96] contains an identical chapter, but extra attention is paid to the engine failure (control and roll angle responses, minimum control speed), the descending flights (throttle responses, energy principles) and windshear ($\mathcal{F}$-factor, throttle responses).

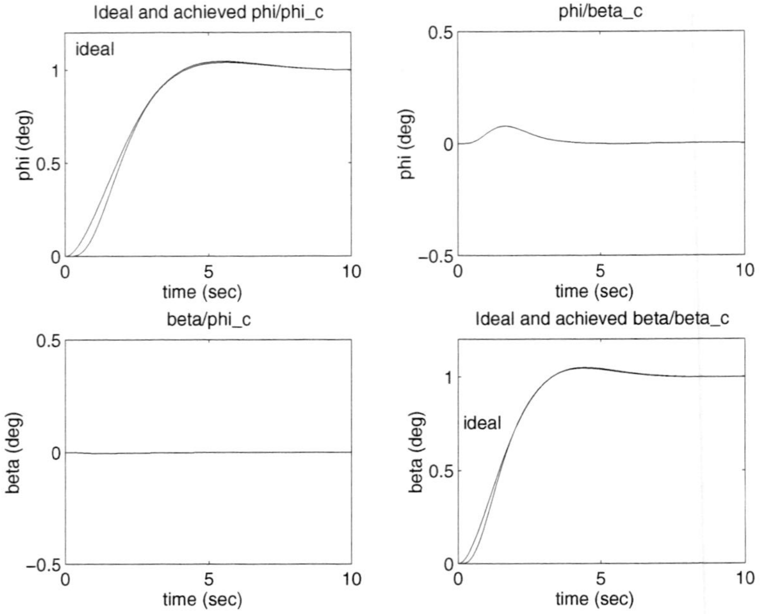


Figure 4.16: Ideal and achieved roll and sideslip angle command responses

## 4.6 Results of the Automated Evaluation Procedure

### 4.6.1 Introduction

This section presents simulation results of the previously designed contollers. The evaluation mission and scenario have been defined in section 4.1. Both 'overall tracking performance' and 'inner-loop behaviour' of the controlled system are evaluated by imposing bounds on key variables during the maneuvre. The performance behaviour of the controlled aircraft is based on four separate flight segments as defined in section 4.1. The numerical results of the evaluation are given in section 4.6.6.
The main robustness demand is taken into account by plotting the three trajectories resulting with nominal, most forward and most aft positions for the centre of gravity.

### 4.6.2 Segment I: the effect of engine failure

As the RCAM aircraft model is twin-engined, an engine failure will mainly result in a lateral deviation from the flight track.
Figure 4.19 gives a top view of the first trajectory segment. The steady-state lateral deviation should be below an acceptable value with minimal or no over-

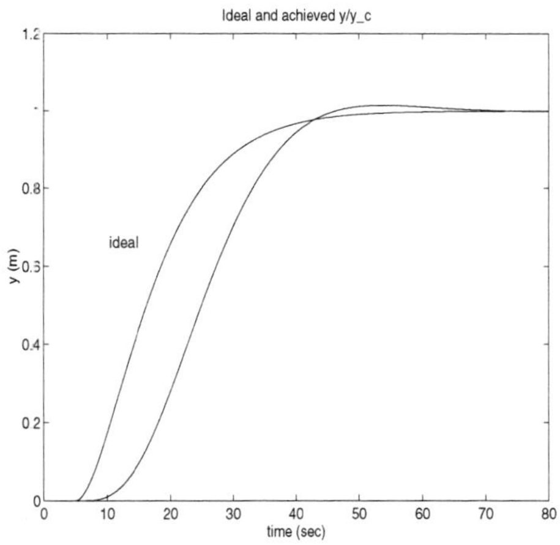


Figure 4.17: Ideal and achieved lateral position command responses

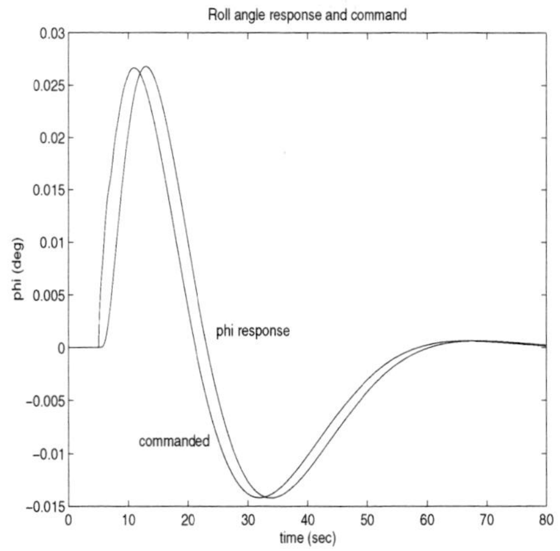


Figure 4.18: Roll command and roll response

shoot. The dashed lines reflect the bounds in which the aircraft should remain during and engine failure. From the time simulation it can be seen that the aircraft performance is satisfactory. Further the aircraft returns on the track with zero sidelsip and a bank angle compensating for thrust asymmetry. Furthermore, the airspeed does not drop below $1.2 \times V_{\text{stall}}$ as one engine is shut off.

An engine failure is characterized by a yawing moment $N_T = -T * y_{CGE}$ resulting from asymmetric thrust $T$ at a distance $y_{CGE}$ from the centre of gravity. We assume that the inoperative engine does not produce any extra drag.
The nondimensional relation for the yawing moment is:

$$C_{N_T} = -\frac{N_T}{\bar{q}Sb} = -\frac{Ty_{CGE}}{\bar{q}Sb} \tag{4.10}$$

If steady level flight is considered the vertical force equilibrium is given by:

$$W = L = C_L \bar{q} S \tag{4.11}$$

With $\bar{q} = \frac{1}{2}\rho V_A^2$ we get:

$$C_{N_T} = -\frac{T}{W} C_L \frac{y_{CGE}}{b} \tag{4.12}$$

Equation 4.12 shows that the lateral yawing moment coefficient due to thrust is proportional to the operative engine thrust $T$, the distance $y_{CGE}$ and the lift coefficient $C_L$ and it is inversely proportional to the airspeed $V_A$.
The equilibrium equations for the side force $Y$, yawing moment $N$ and rolling

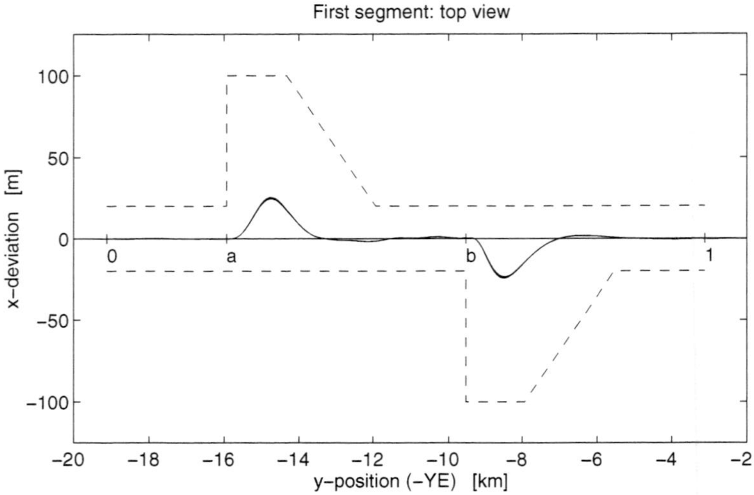


Figure 4.19: Segment I: the effect of engine failure

moment $L$ can be written as follows, assuming $p = r = 0\,\mathrm{deg}$:

$$
\begin{aligned}
C_Y &= C_{y_\beta}\beta + C_{y_{\delta_R}}\delta_R + C_L\mu &&= 0 \\
C_n &= -\frac{T}{W}C_L\frac{y_{CGE}}{b} + C_{n_\beta}\beta + C_{n_{\delta_R}}\delta_R &&= 0 \\
C_l &= C_{l_\beta}\beta + C_{l_{\delta_A}}\delta_A + C_{l_{\delta_R}}\delta_R &&= 0
\end{aligned}
\tag{4.13}
$$

($C_{n_{\delta_A}}$ is assumed to be zero)
For the directional control problem (constant heading) there are in theory infinitly many solutions; there are three equations and four unknowns. In practice a pilot will mostly consider one of the following three limiting solutions:

- zero sideslip angle, this solution is aerodynamically the most effective one, since drag remains along the body $x$-axis;
- no bank angle, this solution needs a somewhat larger rudder solution;
- no rudder deflection, where equilibrium can be achieved only by increasing the sideslip and bank angle.

A controller will probably find a solution in between, unless one of the parameters is explicitly driven to zero in a special engine failure mode or via an integrator in the loop. Since we demanded (approximate) integrator behaviour in the sideslip channel during synthesis of the lateral innerloop controller, we expect $\beta$ to be driven to zero. Therefore we will only consider the zero-sideslip option.
The rudder deflection for this case can be derived from the expression of the yawing moment:

$$
\delta_{R_{ss}} = \frac{\frac{T}{W}C_L\frac{y_{CGE}}{b}}{C_{n_{\delta_R}}}
\tag{4.14}
$$

Some bank angle is needed to provide side force equilibrium. This situation is sketched in figure (4.20).

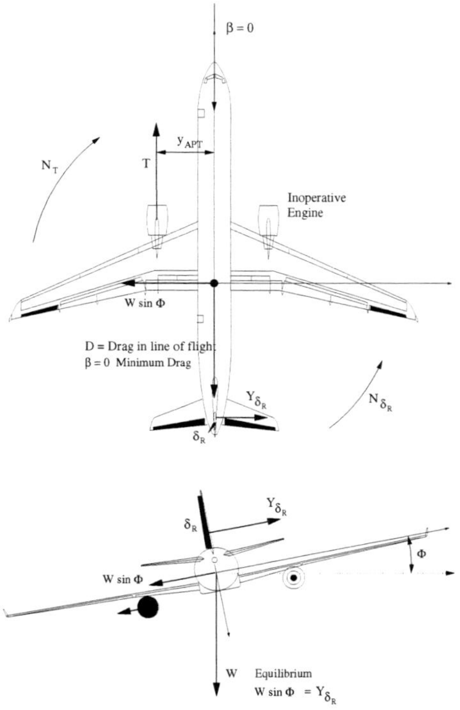


Figure 4.20: Asymmetric equilibrium for zero sideslip

The required bank angle can be obtained from equation 4.13:

$$\mu = \frac{-C_{y_{\delta_R}} \delta_R}{C_L} \tag{4.15}$$

The bank angle $\mu$ has to remain in most of the situations less than 5 degrees. During the RCAM evaluation thrust of the left engine (engine 1) is reduced to $0.10\ 10^5\ N$ (idle), while it is flying at a speed of $V_A = 80\ m/s$. The nominal aircraft mass is 120 000 $kg$. The required thrust for steady horizontal flight for this mass and airspeed appears to be (from trimming the model) $T = 1.79\ 10^5\ N$, so that the operative engine at least will have to deliver a thrust of $(1.79 - 0.10)\ 10^5 = 1.69\ 10^5\ N$. The difference in delivered thrust between both engines is then: $(1.69 - 0.10)\ 10^5 = 1.59\ 10^5\ N$, causing the asymmetrical thrust moment.
The lift coefficient will approximately be equal to ($\rho = \rho_0 = 1.225\ kg/m^3$ and $S = 260\ m^2$):

$$C_L = \frac{W}{1/2 \rho V_A^2 S} = \frac{120000\ 9.81}{1/2\ 1.225\ 80^2\ 260} = 1.16$$

Furthermore, from chapter 2 in [Fli96] $y_{CGE}$ of the operative (right) engine can be obtained: $y_{CGE} = y_{APT2} = 7.94\ m$, $b = 45\ m$ and $C_{n_{\delta_R}} = -0.092$. The required rudder deflection can now be calculated:

$$\delta_{R_{ss}} = \frac{1.59\ 10^5/(120000\ 9.81)\ 1.16\ (7.94/45)}{-0.092} * 57.3 = -17.2\,\text{deg} \quad (4.16)$$

The bank angle will be equal to: (with $C_{Y_{\delta_R}} = 0.24$)

$$\mu = \frac{-0.24\ (-17.2/57.3)}{1.16} * 57.3 = 3.6\,\text{deg} \quad (4.17)$$

This is less than the 5 degrees accepted in practice. If this limit would have been exceeded, another equilibrium option had to be adopted, for example by allowing a small sideslip angle. Also the rudder deflection is less than the maximum of 30 degrees. Finally, we can calculate from eq.4.13 the steady-state aileron deflection: (with $\beta = 0$, $C_{l_{\delta_R}} = 0.032$ and $C_{l_{\delta_A}} = -0.088$)

$$\delta_A = -\frac{C_{l_{\delta_R}}}{C_{l_{\delta_A}}}\delta_R = -\frac{0.032}{-0.088}(-17.2) = -6.3\,\text{deg} \quad (4.18)$$

In figure 4.21 the sideslip angle and roll angle reponses are depicted. $\beta$ gets close to zero, while the roll angle is approximately 3.6 deg. Since for civil aircraft the bank angle is approximately equal to the roll angle, the bank angle will also be about 3.6 deg. Also note that the maximum peak of $\mu = 8.4$ deg is within the 10 deg limit allowed in the specifications. When the failed engine is restarted, the bank angle shows a large overshoot to the opposite side. This is not very surprizing: the response is approximately opposite compared to the situation when the engine is failed. However, in this way the specification of less than 50% bank angle overshoot due to engine restart is violated.
In figure 4.21 also the heading rate response $\dot{\psi}$ can be found: the limit of 3 deg $/s$ is not violated. Finally we will look at the responses of the controls. As expected, we can see from figure 4.22 that the throttle deflection of the left engine doubles. It can also be seen that the steady-state responses of the aileron and rudder deflections match the calculated values reasonably well.

### 4.6.3 Segment II: the 3 *deg/s* turn

At he moment the turn starts, the perfect following of the required trajectory and the desire to perform a coordinated turn would imply a sudden change in the aircraft's bank angle, which is only possible with an infinitely high roll rate. Obviously this is unwanted, such that deviations from the desired trajectory at the start (and the finish) of the turn are unavoidable. To demonstrate this, figure A.4 gives a top view of the trajectory for all three centre of gravity locations plus bounds within which an acceptable performance should lie.
Figure 4.24 provides a closer look at the lateral deviations.
In a steady coordinated turn, the aircraft will find a steady-state bank angle at approximately zero sideslip, see figure 4.25.

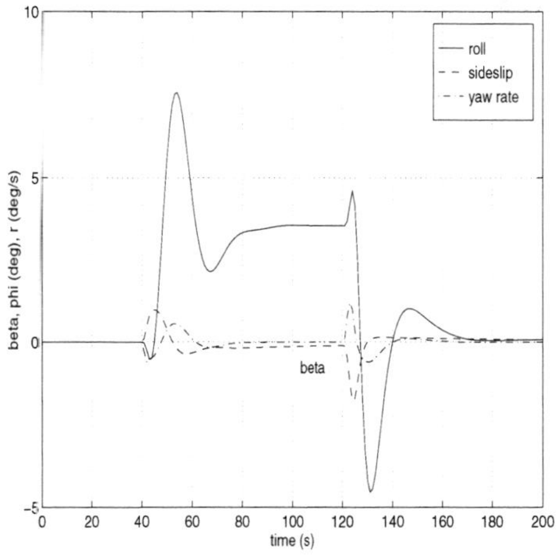


Figure 4.21: Sideslip, roll angle, and heading rate responses during engine failure

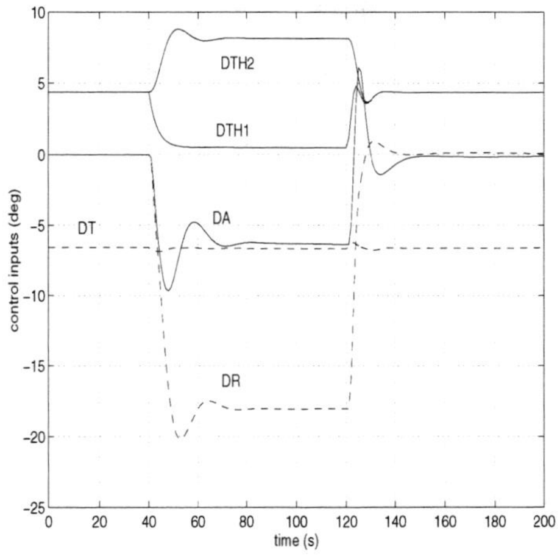


Figure 4.22: Control deflections during engine failure

The bank angle can easily be calculated from the well known formula:

$$\mu = \tan^{-1}\left(\frac{V\dot{\chi}}{g}\right) \tag{4.19}$$

With an airspeed of $V_A = 80\ m/s$ and turning rate of $\dot{\chi} = 3\deg/s$ the required bank angle becomes:

$$\mu = \tan^{-1}\left(\frac{80\ 3/57.3}{9.81}\right) * 57.3 = 23.1\deg \tag{4.20}$$

This will cause the load factor to become:

$$n_z = \frac{1}{\cos\mu} = \frac{1}{\cos 23.1} = 1.1 \tag{4.21}$$

In the coordinated turn the sideforce along the $y_B$-axis must be zero, so that the following relation holds:

$$C_Y = C_{Y_\beta}\beta + C_{Y_{\delta_R}}\delta_R + C_{Y_r}\frac{rb}{2V} = 0 \tag{4.22}$$

Although the aircraft does not really fly a constant bank angle turn, for passenger comfort the lateral acceleration should be as small as possible. In figure 4.26 we can see that sideslip is very close to zero due to integration in the controller, while the bank angle ($\mu$) goes to 23 deg. For equilibrium in the turn, a rudder deflection of approximately $-4$ deg is required, see figure 4.27. Since the sideslip angle is driven to zero, a lateral force remains:

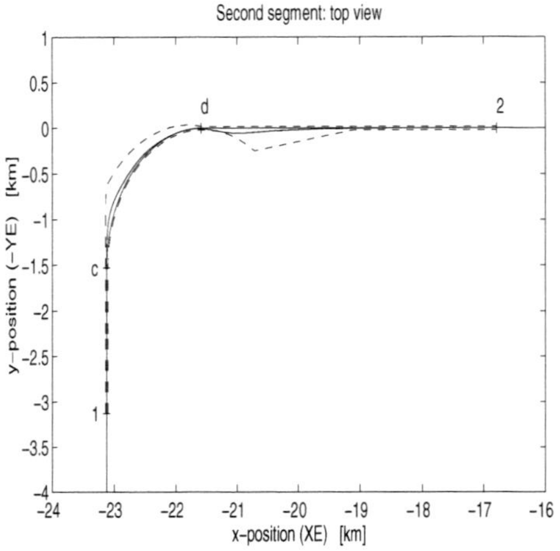


Figure 4.23: Segment II: the 3 *deg/s* turn

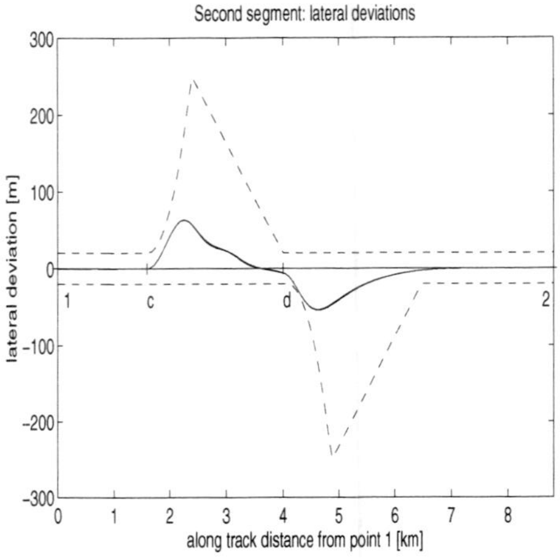


Figure 4.24: Segment II: lateral deviations during the 3 *deg/s* turn

$C_Y = C_{Y_\beta}\beta + C_{Y_{\delta_R}}\delta_R \approx 0.24 * -4/57.3 = -0.017$. Apparently integration of the sideslip prevents accurate turn co-ordination.

In figure 4.27 we can see that the responses are quite oscillatory. This is caused by interaction between the longitudinal and lateral motions. In an improved design this could be adjusted by turn corrections in the longitudinal controller.

The large overshoot in the bank angle in figure 4.26 arises due to the sudden input of a roll feedforward. As a consequence, the maximum roll angle is exceeded by 5 deg when the turn is initiated. The overshoot could be decreased at the cost of the tracking error.

### 4.6.4 Segment III: the capture of the -6 and -3 degrees glideslope

We start with a glideslope of -6 *deg*; again it is unavoidable that the aircraft leaves the desired trajectory. It should return to the trajectory without overshoot and well within a period of 30 *s*. After that we go to a glideslope of -3 *deg* such that we get an inverse behaviour with respect to the desired trajectory, that should be about half the size of the first response (if the system has a more or less linear behaviour).

In figure A.6 the longitudinal response of the aircraft is plotted for three centre of gravity locations and with bounds that specify acceptable behaviour.

The vertical deviations from the desired glideslope are plotted in figure 4.29.

Figure 4.30 shows aircraft and control responses during the descent phase of the trajectory. During this simulation there was no wind. From section 4.5 the

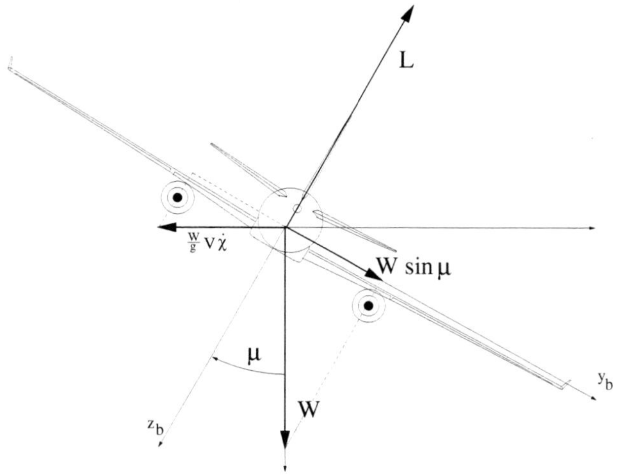


Figure 4.25: Aircraft in coordinated turn.

following equation was derived: (eq. 4.9)

$$\frac{\dot{V}}{g} + \sin\gamma = \frac{T - D}{W} \tag{4.23}$$

Since velocity has to remain constant and also the drag $D$ will not vary too much, we expect the throttle response to be:

$$\sin\gamma = \frac{\Delta T}{W} \tag{4.24}$$

Since $\delta_{TH_i} = T_i/(120000g)$, the throttle position at $m = 120000\ kg$ will be $T/W$. In figure 4.30 can be seen that the throttles indeed follow the required $\gamma_c$-profile. The term $\dot{V}/g$ remains small, while the flightpath angle response closely follows the commanded value. The tailplane shows peaks at the moments the aircraft has to rotate. This device trims the aircraft along the trajectory, while the throttles are moving.

### 4.6.5 Segment IV: the final approach with windshear

While on final approach with a glideslope of -3 *deg* the effect of a windshear is considered. Figure A.10 shows the longitudinal response of the aircraft for three centre of gravity locations and with bounds that specify acceptable behaviour. The vertical deviations from the desired glideslope are plotted in figure A.10. To analyze the effect of windshear, energy considerations are again well applicable. We will start with the force equation in $x$-direction, however this time along the airspeed vector $V_a$, see figure 4.32. The inertial speed can be written as follows:

$$V = V_A + W_{X_E}\cos\gamma_a - W_{Z_E}\sin\gamma_a \tag{4.25}$$

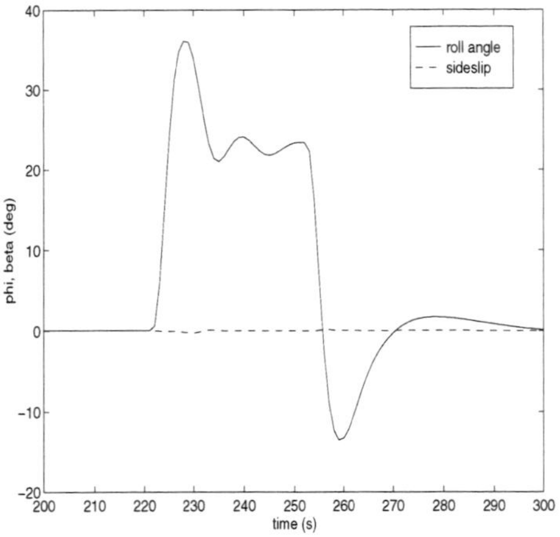


Figure 4.26: Roll and sideslip angle responses during turn

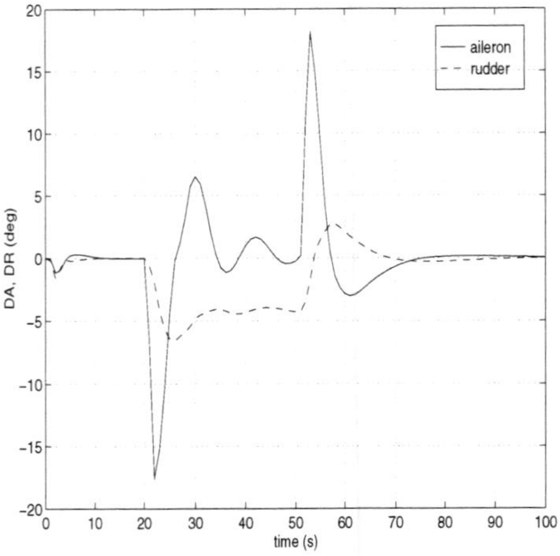


Figure 4.27: Rudder and aileron deflections in turn

We assume here that $\cos(\gamma_a - \gamma) \approx 1$. The force equation along $V_A$ (in $x_A$-direction) is:

$$\dot{V}_A = \frac{T - D}{W} - \dot{W}_{X_E} \cos\gamma_a + \dot{W}_{Z_E} \sin\gamma_a - \sin\gamma_a \tag{4.26}$$

The vertical speed can be expressed as the sum of the vertical velocity of the airmass and the vertical component of the airspeed ($= V_A \sin\gamma_a$):

$$\dot{h} = -W_{Z_E} + V_A \sin\gamma_a \tag{4.27}$$

As an approximation, we now can write $\sin\gamma$ as:

$$\sin\gamma \approx -\frac{W_{Z_E}}{V_A} + \sin\gamma_a \tag{4.28}$$

Substitution of eq. 4.28 in eq. 4.26 and rearranging terms gives us:

$$\frac{\dot{V}_A}{g} + \sin\gamma = \frac{T - D}{W} - \dot{W}_{X_E} \cos\gamma_a + \dot{W}_{Z_E} \sin\gamma_a - \frac{W_{Z_E}}{V_A} \tag{4.29}$$

This again is an expression of a specific energy rate, however the acceleration term is airmass related. The airmass terms on the righthand form the so called $\mathcal{F}$-factor:

$$\mathcal{F} = \dot{W}_{X_E} \cos\gamma_a - \dot{W}_{Z_E} \sin\gamma_a + \frac{W_{Z_E}}{V_A} \tag{4.30}$$

If $\mathcal{F}$ is negative, the specific excess power is enlarged by the wind terms, thus there is a performance increasing effect. For positive $\mathcal{F}$ the opposite occurs.

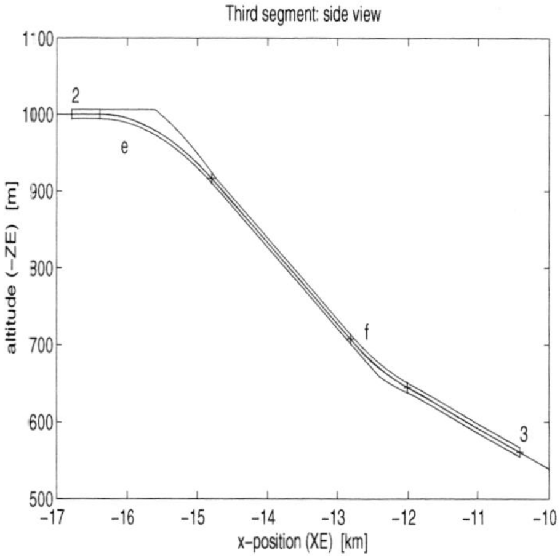


Figure 4.28: Segment III: the capture of the -6 and -3 degrees glideslope

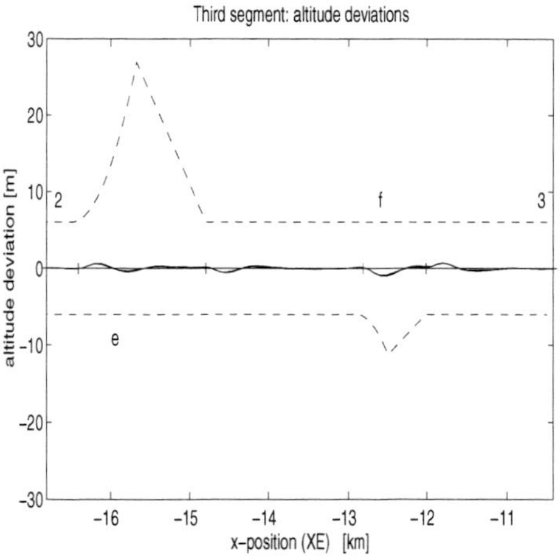


Figure 4.29: Segment III: vertical deviations from the desired glideslope

In figure 4.33 the windshear $\mathcal{F}$-factor and its components during the RCAM evaluation is given (nominal configuration, no gust).
It is clear that the vertical wind and the horizontal shear have a dramatical effect on the performance. In the same figure the throttle response has been plotted. It is interesting to see that the longitudinal controller (that controls inertial flightpath and airspeed) tries to compensate for the loss of power due to the disturbance. This again illustrates that the controller uses throttles in the way a pilot tends to use them: as an energy input. Due to inertia of the aircraft body, the airspeed still varies, as can be seen in figure 4.34. The flightpath angle deviation remains relatively small.
The windshear encountered along the flightpath clearly is not beyond the performance limits of RCAM [LB96]. In fig. 4.33 we see that thrust is initially decreased. In severe windshear this is very undesirable, since in this phase the aircraft should increase its energy level in order to fly through the core of the windfield without loosing too much airspeed and altitude.

### 4.6.6 Numerical Results

Here a table of numerical results based on the discussed simulation results is given. For the motivation and calculation principle of the various results see appendix A.
The overall evaluation results shown in last column of table 4.2 clearly indicate that the design objective has been met. Remark that the comfort requirement due to a violation of the normal acceleration bound has been overridden in the segments II and III. This has been the case for the majority of the RCAM designs reported in [MBT97].

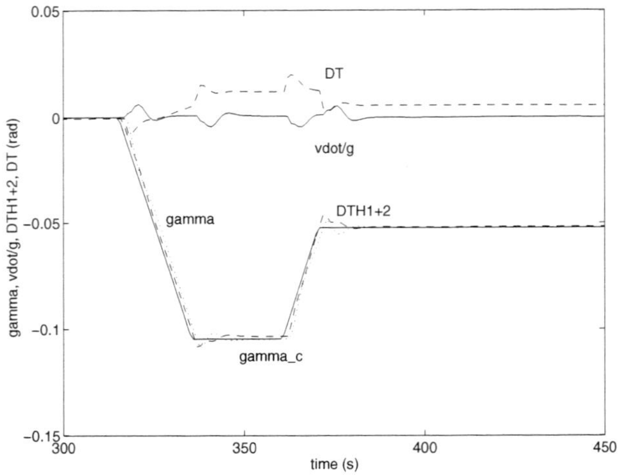


Figure 4.30: Flightpath, acceleration and control response during descent

| | Segment I | Segment II | Segment III | Segment IV | Total |
|---|---|---|---|---|---|
| Performance | 0.1305 | 0.1626 | 0.0511 | 0.0903 | 0.1086 |
| Robustness | 0.0293 | 0.0215 | 0.0371 | 0.0837 | 0.0429 |
| Comfort | 0.4204 | 1.1386 | 1.6027 | 0.6393 | 0.9502 |
| Safety | 0.0048 | 0.1038 | 0.0109 | 0.0448 | 0.0411 |
| Power | 0.0042 | 0.0058 | 0.0152 | 0.0302 | 0.0138 |

Table 4.2: Numerical results of the evaluation procedure

## 4.7 Conclusions

In the previous sections the application of $\mu$-synthesis for a flight control system has been discussed. The nonlinear evaluation has shown that the performance of the designed controller is satisfactory in terms of tracking and decoupled reponses.

To demonstrate how the $\mu$-method can cope with uncertainties we have avoided gain scheduling along the varying parameters and mode switching for configuration change. To keep the problem simple we designed with general input-multiplicative uncertainty models, which appeared to be sufficient. This has been established from a detailed parametric robust stability analysis using $\mu$. The results of this study have been summarized in chapter 35 of [MBT97]. The great feature of the methodology is that we can account simultaneously for uncertainties in multiple specific parameters (e.g. mass, $CG_x$, transport delay). The $\mu$ analysis gives us bounds on the magnitudes of the uncertainties which can be tolerated until instability occurs. To this end an LFT model reflecting mass, centre of gravity and transport delay variations has been developed. The

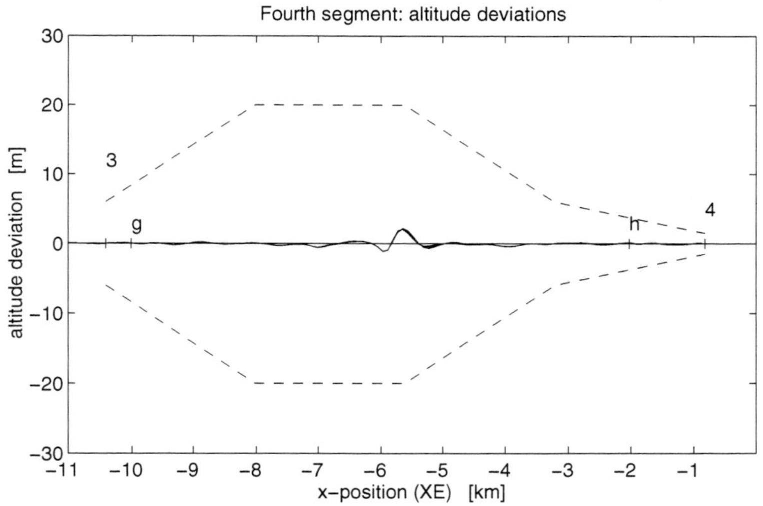


Figure 4.31: Segment IV: vertical deviations from the desired glideslope

$\mu$ value indicating the actual parametric stability robustness level is $\mu = 0.49$. The value of $\mu = 0.49$ is very close to the robust stability level achieved in our design using the general input-multiplicative uncertainty models. In that respect our design was not unnecessarily complicated by the tedious LFT model generation reported in [MBT97, LMVB96, VLMG96]. Nevertheless, the parametric LFT model is very well suited for a more refined design. Such a model will allow to put all the design effort in fine tuning the trade-off between the various performance requirements. But before trying to get better control solutions it is recommended to perform detailed flight simulator studies in order to capture a refined set of design specifications.

One shortcoming of the method it that the resulting controllers are of the order of the scaled interconnection structure. Although commercially available model reduction techniques give some relief these do not provide satisfactory reduction levels to be acceptable for industrial applications as this one. However, much effort has been made and encouraging results have been provided by Packard etal [Pac95]. The reduction problem has been approached from a $\mu$ perspective and incorporated in the original $\mu$-synthesis problem. The reduction levels reported are doubled without loosing the initially achieved performance. For our problem this would mean that it is possible to easiliy obtain sixth to eighth order controllers with the proposed technique which is in industry an acceptable controller size.

Beyond $\mu$, a lot of progress has been made in the robust control area. The potentials of using *Linear Matrix Inequalities* (LMI's) for controller analysis and synthesis has been widely recognized. Nowadays there are several LMI

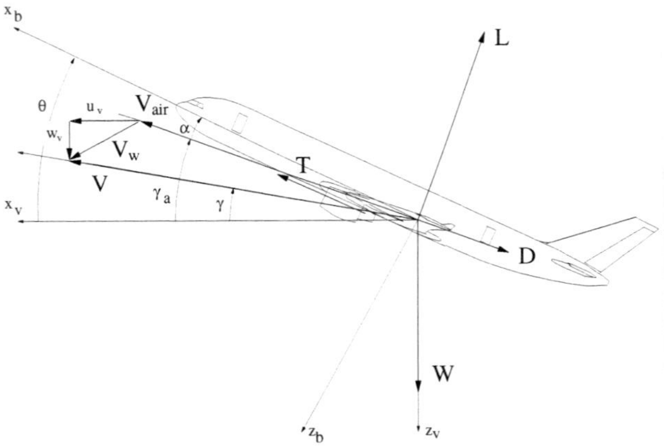


Figure 4.32: Longitudinal aircraft motion in a windfield.

formulations for the synthesis of robust gain-scheduled controllers which take into account the time varying nature of the plant [Wu95, AA96, Hel95, BWS98, NSB96, Sch95b]. Illustrating the practical relevance of those advanced methods is the objective in the further development of this thesis.

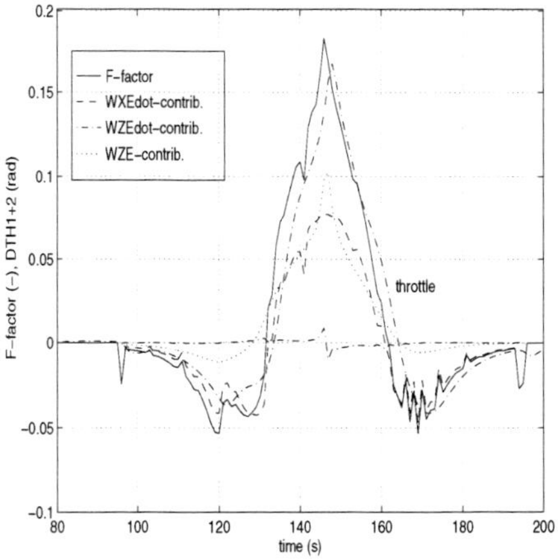


Figure 4.33: F-factor of encountered windshear

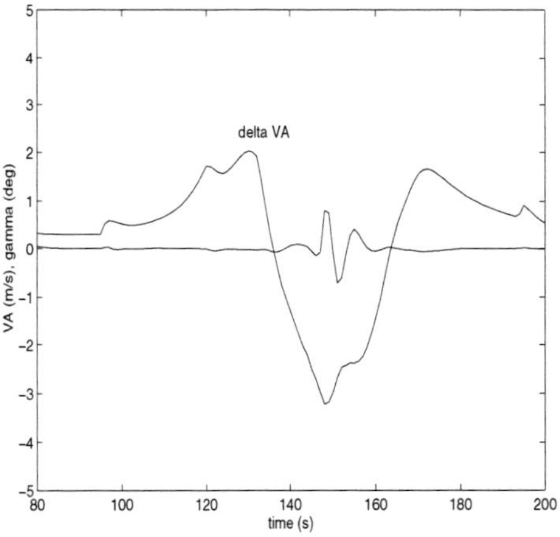


Figure 4.34: Airspeed and flight-path response during windshear

# Chapter 5

# Robust LPV Control with Bounded Rates: Missile LPV-Design

*This chapter discusses an approach to the control of Linear Parametrically Varying (LPV) systems that can take the rate of parameter variations into account and also guarantees robustness against parametric and dynamic uncertainties. To illustrate the technique we consider a missile control problem that has been extensively studied in the literature. For this nonlinear model, the objective is to design a controller with guaranteed performance robustness over a given operating range.*

## 5.1 Introduction

The classical approach to gain scheduling relies on the interpolation of controllers designed for frozen parameters as, e.g., the operating conditions. This procedure, even if seemingly working well in practice, does not take the time-variations of the involved parameters into account. It has been shown in [SA92] that one cannot provide a priori stability and performance guarantees for linear systems that depend on a time-varying parameter by using results for Linear Time Invariant (LTI) systems.

Therefore, one has to base system analysis and controller synthesis on results that are dedicated to Linear Parametrically Varying (LPV) systems.

In the early nineties Packard, Becker and in parallel Apkarian and Gahinet [Pac94, Bec93, GA95] have shown how the robust control framework for LTI systems can be generalized to arrive at a systematic design procedure for LPV systems.

The desired stability and performance guarantees are provided under the conservative assumption that the rate of change of the parameters is unbounded.

Hence, a refined approach similar to the one proposed in [Wu95, AA96, Wil96, Sch95b, Bec96, LS97, Lee97] will be pursued in this chapter taking explicitely into account the bounds on the parameter rate of change.
In addition, we incorporate robustness against structured uncertainties in the design procedure. In contrast to [BP94, AA96, LS97], we employ the more general framework of Integral Quadratic Constraints (IQC's) [RM94] to reflect the nature of the uncertainties.
The chapter is structured as follows.
First in section 5.2, the description of uncertain LPV systems is given. Then, it is shown how structural knowledge about parameter variations and uncertainties are cast into suitable constraints for multipliers or scalings. This leads to the analysis characterization of stability and robustness of uncertain LPV systems in terms of a Scaled Differential Bounded Real Lemma (SDBRL). In the corresponding LPV synthesis problem, the to be designed controller parameters render the inequalities non-convex. We show that the controller parameters can be transformed into new variables that enter the inequalities in an affine fashion [MOS95, Sch95b]. For fixed scalings we end up again with differential linear matrix inequalities. Introducing basis functions and gridding the parameter set will result, in finitely many linear matrix inequalities that can be readily solved [GNLC95]. If optimizing as well over the scalings, we have to resort to a $D$-$K$-like iteration [BDG+93b].
The theory is illustrated in section 5.3 on a missile benchmark problem as studied in [NRR93] and [Wu95]. In contrast to [Wu95], which considers the nominal performance LPV problem, we also address robust performance issues. Starting from a given nonlinear model, we obtain through analytical manipulations similar to [BP92, Hel95] an uncertain LPV representation accessible for design. Note that, for this model it is straightforward to obtain a suitable LPV representation. In general, however it is not obvious how to arrive at such a description from input output data. For some results on the identification of LPV systems we refer to [Lee97]. Then, the design specifications are translated into suitable weighting functions as in the $H_\infty$-approach. We end up with an interconnection structure and perform the iteration that is comparable to the $D$-$K$ procedure in $\mu$-synthesis [BDG+93b]. Finally, we validate in section 5.4 the robust LPV controller by nonlinear simulations.

## 5.2 LPV design

### 5.2.1 LPV systems

The uncertain LPV system is described by

$$\begin{bmatrix} \dot{x} \\ z \\ y \end{bmatrix} = \begin{bmatrix} A(p) & G(p) & B(p) \\ H(p) & F(p) & E(p) \\ C(p) & D(p) & \end{bmatrix} \begin{bmatrix} x \\ w \\ u \end{bmatrix} \tag{5.1}$$

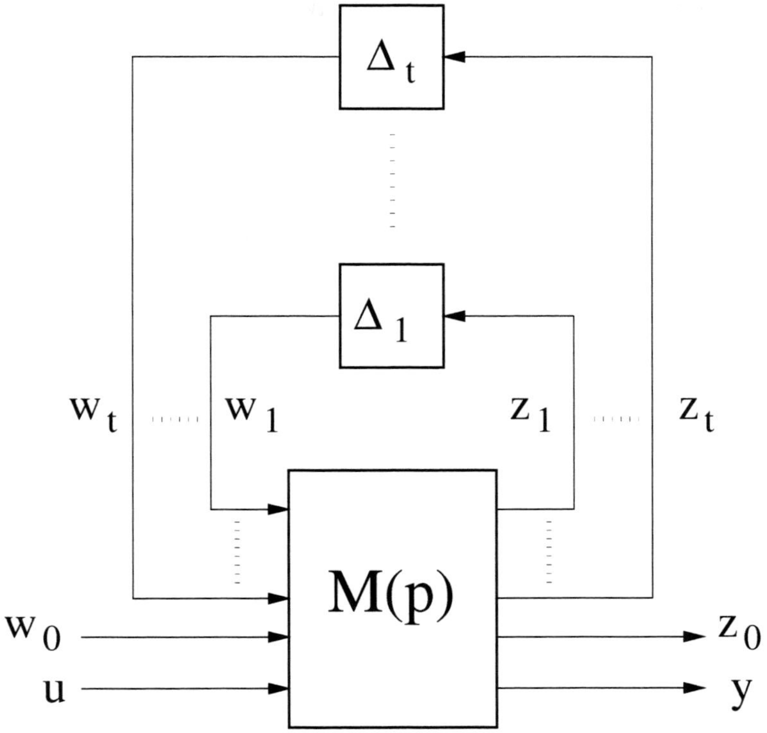


Figure 5.1: LPV system with uncertainty and performance channels

where, with a suitable partition of the signals $w = [w_0, w_1, \ldots, w_k]^T$ and $z = [z_0, z_1, \ldots, z_k]^T$, the uncertainty enters as

$$w_i = \Delta_i z_i, \quad i = 1, \ldots, k, \tag{5.2}$$

and $w_0 \mapsto z_0$ is the channel to describe the performance specification (figure 5.1). The parameter $p(t)$ and its rate of variation $\dot{p}(t)$ are assumed to be contained in the a priori given compact sets $P$ and $P_r$ respectively.
For controller design the parameter $p(t)$ is assumed to be on-line measurable. Hence, LPV controllers take the form

$$\begin{bmatrix} \dot{x}_c \\ u \end{bmatrix} = \begin{bmatrix} A_c(p) & B_c(p) \\ C_c(p) & D_c(p) \end{bmatrix} \begin{bmatrix} x_c \\ y \end{bmatrix} \tag{5.3}$$

such that the resulting closed-loop system is described by

$$\begin{bmatrix} \dot{\xi} \\ z \end{bmatrix} = \begin{bmatrix} \mathcal{A}(p) & \mathcal{G}(p) \\ \mathcal{H}(p) & \mathcal{F}(p) \end{bmatrix} \begin{bmatrix} \xi \\ w \end{bmatrix} \tag{5.4}$$

together with (5.2), where the individual matrix elements are:

$$\begin{aligned} \mathcal{A}(p) &= \begin{bmatrix} A(p)+B(p)D_c(p)C(p) & B(p)C_c(p) \\ B_c(p)C(p) & A_c(p) \end{bmatrix}, \\ \mathcal{G}(p) &= \begin{bmatrix} G(p)+B(p)D_c(p)D(p) \\ B_c(p)D(p) \end{bmatrix}, \\ \mathcal{H}(p) &= \begin{bmatrix} H(p)+E(p)D_c(p)C(p) & E(p)C_c(p) \end{bmatrix}, \\ \mathcal{F}(p) &= \begin{bmatrix} F(p)+E(p)D_c(p)D(p) \end{bmatrix}. \end{aligned}$$

### 5.2.2 Analysis of uncertain LPV systems

In this section an analysis result is given that provides robust stability and robust performance conditions for uncertain LPV systems given by (5.4) and (5.2). For that purpose we introduce scalings that characterize the nature of the uncertainties $\Delta_i$ affecting the plant. The uncertainties satisfy certain integral quadratic constraints as proposed by [RM94]. Define for each channel a collection of scaling matrices $Q_i$, $S_i$, $R_i$ such that the IQC

$$\int_0^T \begin{bmatrix} w_i \\ z_i \end{bmatrix}^T \begin{bmatrix} Q_i & S_i^T \\ S_i & R_i \end{bmatrix} \begin{bmatrix} w_i \\ z_i \end{bmatrix} dt \geq 0 \tag{5.5}$$

holds for $T \geq 0$ and for all signals $w_i$, $z_i$ of finite energy that are related by $w_i = \Delta_i z_i$.

- time-varying parametric uncertainties $w_i(t) = [\delta_i(t)I]z_i(t)$, $|\delta_i(t)| \leq 1$:
$$Q_i < 0 \;\; R_i = -Q_i, \;\; S_i + S_i^T = 0.$$
- dynamic uncertainty $\Delta_j : L_2[0,\infty) \mapsto L_2[0,\infty)$ with gain not larger than 1:
$$Q_i = q_i I < 0, \;\; R_i = -Q_i, \;\; S_i = 0.$$

The closed-loop system performance is measured via the $L_2$-gain of the channel $w_0 \to z_0$. The $L_2$-gain of this channel is bounded by the value $\gamma$ if the IQC (5.5) holds with the fixed scalings

$$Q_0 = -\gamma I, \;\; R_0 = \frac{1}{\gamma} I, \;\; S_0 = 0.$$

Finally, the scalings are collected into block-diagonal matrices as
$Q = \mathrm{diag}(Q_0, Q_1, \ldots, Q_k)$, $R = \mathrm{diag}(R_0, R_1, \ldots, R_k)$, and
$S = \mathrm{diag}(S_0, S_1, \ldots, S_k)$.
Using this setup, robust stability and robust performance for uncertain LPV systems are achieved under the solvability conditions given by the scaled differential Bounded Real Lemma. The proof of this Lemma can be found in [Hel95, SG95, Sch95b, Wil96]. For notational convenience define $\mathrm{sy}(M) := M + M^T$.

**Theorem 5.2.1** *Suppose there exist smooth and bounded functions $\mathcal{X}(p)$, $Q(p)$, $R(p)$, $S(p)$ on $P$ such that*

$$\mathcal{X}(p) > 0$$

$$\begin{bmatrix} \mathrm{sy}(\mathcal{X}(p)\mathcal{A}(p)) + \mathcal{X}'(p,\dot{p})\,\mathcal{X}(p)\mathcal{G}(p) + (S(p)\mathcal{H}(p))^T & * & \\ \mathcal{G}^T(p)\mathcal{X}(p) + S(p)\mathcal{H}(p) & \mathrm{sy}(S(p)\mathcal{F}(p)) + Q(p) & * \\ R(p)\mathcal{H}(p) & R(p)\mathcal{F}(p) & -R(p) \end{bmatrix} < 0 \tag{5.6}$$

*holds for all $p \in P$ and $\dot{p} \in P_r$. Then, for all parameter curves $(p(t), \dot{p}(t)) \in P \times P_r$ and for all uncertainties (5.2), the system (5.4) remains stable and the $L_2$-gain of the performance channel is bounded by $\gamma$.*

Here $\mathcal{X}'(p,\dot{p})$ is defined as

$$\mathcal{X}'(p,\dot{p}) = \sum_{j=1}^{m} \frac{\partial \mathcal{X}}{\partial p_i}(p)\dot{p}_i.$$

Hence we have to find a parameter dependent Lyapunov function and parameter dependent scalings to satisfy a differential linear matrix inequality [AA96, Sch95b]. This generalizes the idea of using a constant Lyapunov function [Bec93, GA95] and constant scalings for arbitrarily fast varying parameters.
To solve the inequalities numerically we choose continuously differentiable functions $f_1(p) \ldots f_l(p)$ and search for the coefficients in the expansion

$$\begin{bmatrix} \mathcal{X}(p) \\ Q(p) \\ R(p) \\ S(p) \end{bmatrix}^T = \sum_{j=1}^{l} f_j(p) \begin{bmatrix} \mathcal{X}_j \\ Q_j \\ R_j \\ S_j \end{bmatrix}^T .$$

The resulting infinite number of LMIs are reduced to finitely many inequalities by picking a finite number of points in $P$ and $P_r$. If $P_r$ is described as a convex combination of finitely many vertices, it suffices to choose the extreme points since the parameter $\dot{p}$ appears linearly in (5.6) [Wu95, AA96].

### 5.2.3 Controller synthesis

The synthesis problem consists of designing a controller (5.3) that minimizes the robust performance level $\gamma$ as characterized in Theorem 5.2.1. However, the inequalities (5.6) are not linear in all the unknowns, the Lyapunov function, the scalings and the controller parameters. It has been shown in [MOS95, Sch95b, SGC97] how the synthesis inequalities, can be linearized, for fixed scalings through a suitable nonlinear transformation $\mathcal{X}(p)$ of the controller parameters $A_c$, $B_c$, $C_c$ and $D_c$. The transformation matrix $\mathcal{X}(p)$ and its reciprocal $\mathcal{X}(p)^{-1}$ are respectively given by $\begin{bmatrix} X(p) & U(p) \\ U^T(p) & * \end{bmatrix}$ and $\begin{bmatrix} Y(p) & V(p) \\ V^T(p) & * \end{bmatrix}$. The new controller parameters in which the synthesis inequalities are linear are given by:

$$\begin{aligned}
K(p,\dot{p}) &= X(p)[A(p) + B(p)D_c(p)C(p)]Y(p) + \\
&\quad + U(p)B_c(p)C(p)Y(p) + X(p)B(p)C_c(p)V^T(p) + \\
&\quad + U(p)A_c(p)V^T(p) + X'(p,\dot{p})Y(p) + U'(p,\dot{p})V^T(p) \\
L(p) &= X(p)B(p)D_c(p) + U(p)B_c(p) \\
M(p) &= D_c(p)C(p)Y(p) + C_c(p)V^T(p) \\
N(p) &= D_c(p).
\end{aligned} \tag{5.7}$$

Then synthesis inequalities are simply obtained by performing the substitutions

$$\mathcal{X}\mathcal{A} \to \begin{bmatrix} XA + LC & K \\ A + BNC & AY + BM \end{bmatrix}$$

$$\mathcal{X}\mathcal{G} \to \begin{bmatrix} XG + LD \\ G + BND \end{bmatrix}, \quad \mathcal{H}^T \to \begin{bmatrix} (H + ENC)^T \\ (HY + EM)^T \end{bmatrix}$$

$$\mathcal{X} \to \begin{bmatrix} X & I \\ I & Y \end{bmatrix}, \quad \mathcal{X}' \to \begin{bmatrix} X' & 0 \\ 0 & -Y' \end{bmatrix}$$

in (5.6). The dependence on $p$ and $\dot{p}$ has been dropped for notational simplicity. As in the analysis case the solution of the synthesis inequalities requires the choice of suitable basis functions and a grid on the parameter space. However, the resulting inequalities are still nonlinear in the new variables and the scalings together. Hence, we have to resort to a $D$-$K$-like iteration scheme that proceeds as follows: Start with the uncertainty scalings $Q_i(p) = -I$, $R_i(p) = I$, $S_i(p) = 0$ and iterate the following two steps until the performance level $\gamma$ cannot be improved:

1. Fix the scalings and minimize $\gamma$ over $X(p)$, $Y(p)$ and the transformed controller parameters $K(p,\dot{p})$, $L(p)$, $M(p)$, $N(p)$.

2. Fix the controller parameters $K(p,\dot{p})$, $L(p)$, $M(p)$, $N(p)$ and minimize the performance level $\gamma$ over $\mathcal{X}(p)$ and the uncertainty scalings $Q_i(p)$, $R_i(p)$, $S_i(p)$ as described for analysis.

Suppose the iteration stops with $X(p)$, $Y(p)$, $K(p,\dot{p})$, $L(p)$, $M(p)$, $N(p)$. Then one simply needs to choose nonsingular smooth and bounded functions $U(p)$, $V(p)$ satisfying

$$Y(p)X(p) + V(p)U^T(p) = I \tag{5.8}$$

in order to calculate the controller by solving (5.7) for $A_c(p,\dot{p})$, $B_c(p)$, $C_c(p)$, $D_c(p)$.

Since $A_c(p,\dot{p})$ depends on $\dot{p}$, one needs to measure not only the parameter value $p(t)$ itself but also its rate of variation $\dot{p}(t)$ to implement the resulting controller.

To avoid this undesired structure, we choose $K(p)$ independent of $\dot{p}$ and either $X(p)$ or $Y(p)$ independent of $p$ similarly to [AA96, Bec96]. Exploiting the freedom in the choice of $U(p)$ and $V(p)$ allows to construct a controller that depends on $p$ only:

- If $X(p)$ is parameter dependent and $Y$ is constant, choose $U(p) = I - X(p)Y$ and $V = I$. Differentiating (5.8) reveals

$$YX'(p,\dot{p}) + VU'(p,\dot{p})^T = 0 \tag{5.9}$$

  so that the terms in (5.7) that depend on $\dot{p}$ indeed drop out.

- If $X$ is constant and $Y(p)$ is parameter dependent, choose $U = I$ and $V(p) = I - Y(p)X$. This implies $X'(p,\dot{p}) = 0$, $U'(p,\dot{p}) = 0$ such that, again, the variable $\dot{p}$ disappears in (5.7).

Note that this restriction to constant $X(p)$ or $Y(p)$ certainly introduces conservatism, with the benefit of a simpler controller implementation.
To further speed up the computation, we finally remark that we performed all calculations after eliminating the transformed controller parameters along standard lines by using the projection lemma, and by using scalings with $S(p) = 0$ [AA96]. If columns of $K_y$ and $K_u$ form bases of the the nullspaces of $\begin{bmatrix} B^T & 0 & E^T \end{bmatrix}$ and $\begin{bmatrix} C & D & 0 \end{bmatrix}$ respectively, the reduced synthesis inequalities then read as

$$\begin{bmatrix} X & I \\ I & Y \end{bmatrix} > 0, \tag{5.10}$$

$$K_y^T \begin{bmatrix} \text{sy}(XA) + X' & XG & H^T \\ G^T X & Q & F^T \\ H & F & -R^{-1} \end{bmatrix} K_y < 0, \tag{5.11}$$

$$K_u^T \begin{bmatrix} \text{sy}(AY) - Y' & G & YH^T \\ G^T & Q & F^T \\ HY & F & -R^{-1} \end{bmatrix} K_u < 0. \tag{5.12}$$

The iteration scheme proceeds as follows:

Step 1: Start the iteration without scalings in the uncertainty channels.

Step 2: For fixed scalings $Q$ and $R$, minimize $\gamma$ (in the scaling of the performance channel) over the constraint LMIs (5.10)-(5.12). Reconstruct the controller parameters.

Step 3: For fixed controller parameters, minimize $\gamma$ over the scalings $Q$ and $R$ (with $Q = -R$) using equation (5.6). Go to step 2.

The overall numerical approach is similar to the one proposed by [AA96].

## 5.3 The Missile Control Problem

As an application we have chosen a missile benchmark problem that has been extensively studied in [BP92, NRR93, Wu95, Hel95] and is particularly suited for addressing gain scheduling as well as robustness issues. The objective is to design a longitudinal autopilot for a tail-fin controlled missile providing normal acceleration tracking over a large range of speed and angle-of-attack. In order to arrive at the design model for analysis and synthesis given in Section 5.3.3, the problem specifications given in Section 5.3.2 are translated into suitable weighting filters and uncertainty models. These are based on the LPV missile model what is described in Section 5.3.1.

### 5.3.1 The Missile Model

The nonlinear state equations of the missile are

$$\dot{\alpha} = f_1(\alpha, q, \delta, M) = \frac{\cos(\alpha)^2}{mu} F_z(\alpha, \delta, M) + q \tag{5.13}$$

$$\dot{q} = f_2(\alpha, q, \delta, M) = \frac{M_y}{I_y}(\alpha, \delta, M) \tag{5.14}$$

The aerodynamic nonlinearity and parameter dependence in the missile model are reflected in the normal force and moment coefficients $C_n(\alpha, \delta, M)$ and $C_m(\alpha, \delta, M)$. Taking the missile symmetry into acount, it suffices to consider the positive values of the angle-of-attack. These aerodymanic coefficients are given by the following polynomial expressions:

$$C_n(\alpha, \delta, M) = a_n\alpha^3 + b_n\alpha^2 + c_n(2 + \tfrac{M}{3})\alpha + d_n\delta$$
$$C_m(\alpha, \delta, M) = a_m\alpha^3 + b_m\alpha^2 - c_m(7 - \tfrac{8M}{3})\alpha + d_m\delta$$

where the polynomial coefficients are

$$\begin{aligned} a_n &= +0.000103 \text{ deg}^{-3} & a_m &= +0.000215 \text{ deg}^{-3} \\ b_n &= -0.009450 \text{ deg}^{-2} & b_m &= -0.019500 \text{ deg}^{-2} \\ c_n &= -0.169600 \text{ deg}^{-1} & c_m &= +0.051000 \text{ deg}^{-1} \\ d_n &= -0.034000 \text{ deg}^{-1} & d_m &= -0.206000 \text{ deg}^{-1}. \end{aligned}$$

The coefficients need first to be transformed into radians to comply with the equations of motion (5.13)-(5.14). This aerodynamic model is valid for the missile traveling between Mach = 2 and Mach = 4 at an altitude of 20,000 ft. Typical maneuvers for this missile result in angle-of-attack values ranging between −20 and +20 degrees. Hence the approximation $\cos(\alpha) \approx 1$ is legitimate. Equation (5.13) simplifies to

$$\dot{\alpha} = \frac{F_z(\alpha, \delta, M)}{mV} + q. \tag{5.15}$$

One way to obtain an LPV model for the missile is to parametrize the set of all equilibrium models. For any angle-of-attack $\alpha \in [0, 20]$ $(deg)$ and Mach number

$M \in [2,\ 4]$, the fin deflection and pitch rate

$$\begin{aligned} \delta(\alpha, M) &= -\frac{1}{d_m}\left[a_m\alpha^3 + b_m\alpha^2 - c_m(7 - \frac{8M}{3})\alpha\right] \\ q(\alpha, M) &= -\frac{F_z(\alpha, \delta, M)}{mV} \end{aligned}$$

lead to an equilibrium of equations (5.13)-(5.14). The specific normal force $n_z$ is measured by an accelerometer placed at the center of gravity of the missile. It is defined as $n_z = \frac{F_z}{W}$ where $W = mg$. For convenience we use the shorthands $K_\alpha = \frac{0.7p_0S}{m\ ss}$, $K_q = \frac{0.7p_0Sd}{I_y}$, $K_n = \frac{\pi}{180}\frac{0.7p_0S}{W}$. The Jacobi linearization of the missile dynamics is then given as

$$\begin{bmatrix} \dot{\alpha} \\ \dot{q} \\ \hline q \\ n_z \end{bmatrix} = \left[\begin{array}{c|c} A(\alpha, M) & B(\alpha, M) \\ \hline C(\alpha, M) & D(\alpha, M) \end{array}\right] \begin{bmatrix} \alpha \\ q \\ \hline \delta \end{bmatrix}$$

where

$$\begin{aligned} A &= \begin{pmatrix} K_\alpha M\frac{\partial C_n}{\partial \alpha} & 1 \\ K_q M^2\frac{\partial C_m}{\partial \alpha} & 0 \end{pmatrix} \\ B &= \begin{pmatrix} K_\alpha M\frac{\partial C_n}{\partial \delta} \\ K_q M^2\frac{\partial C_m}{\partial \delta} \end{pmatrix} \\ C &= \begin{pmatrix} 0 & 1 \\ K_n M^2\frac{\partial C_n}{\partial \alpha} & 0 \end{pmatrix} \\ D &= \begin{pmatrix} 0 \\ K_n M^2\frac{\partial C_n}{\partial \alpha} \end{pmatrix}. \end{aligned}$$

These dynamics represent in all equilibrium points given by $(\delta(\alpha, M), q(\alpha, M))$ the family of all linearized systems parametrized by $p = (\alpha, M)$. For a particular parameter value $p$ in the allowable parameter set the LPV dynamics are called frozen and reflect a local linearization of the missile dynamics.

## 5.3.2 Specifications

The specifications to be achieved by the controller have to hold over the whole Mach range $M \in [2,\ 4]$. Therefore, the system should globally provide normal acceleration command tracking features, with rise-time not greater than 0.35 s, overshoot not greater than 10 %, and steady state error not greater than 1 %. The measurements available for control are the normal acceleration $n_z$, the pitch rate $q$ and the Mach number $M$. During a typical maneuver the angle-of-attack ranges within $|\alpha| \leq 20$ degrees, the tail-fin deflection rate should not exceed 25 deg /s per commanded g-level.

As very strong simplifications in the missile modeling have been made, we take the robustness issue originating from the uncertainty in the aerodynamic

coefficients $C_n$ and $C_m$ into account. The uncertainity levels considered are $\Delta C_n = \pm 10$ % and $\Delta C_m = \pm 25$ %.
The controller provides fin commands $\delta_c$ that are processed through second order actuator dynamics given by $G_{\text{act}}(s) = \frac{\omega_a^2}{s^2+2\zeta\omega_a s+\omega_a^2}$, with natural frequency $\omega_a = 150$ rad/s and damping $\zeta = 0.7$. To avoid excitation of unmodeled high frequency dynamics a multiplicative input uncertainty weighting $W_{\text{in}}(s) = 1.5\frac{s+2}{s+80}$ is placed at the actuator.

### 5.3.3 Control Strategy

To realize the specifications over the prescribed operating range, the missile dynamics are reformulated into an uncertain parameter varying system representation. The LPV system with parameter $p = (\alpha, M)$ has an uncertain part arising from the perturbations in the aerodynamic coefficients. Furthermore, we view the angle-of-attack in the parameter vector $p$ as uncertain. This strategy has been chosen to illustrate on the one hand, how uncertainties of various origins can be incorporated in the method and, on the other hand, because of the physical reason that the angle-of-attack $\alpha$ is not measurable. This is in contrast to the less conservative approach shown in [Wu95] where the angle-of-attack $\alpha$ has been used as schedule measurement during the synthesis and implemented via an observer structure.
The angle-of-attack $\alpha$ and the uncertainties in the aerodynamic coefficients $C_m$ and $C_n$ are pulled out of the system and rescaled to $[-1, +1]$. The final uncertainty structure in the missile dynamics is then $\Delta_u = \text{diag}\,(\delta_\alpha I_2,\ \Delta_{C_n},\ \Delta_{C_m})$.

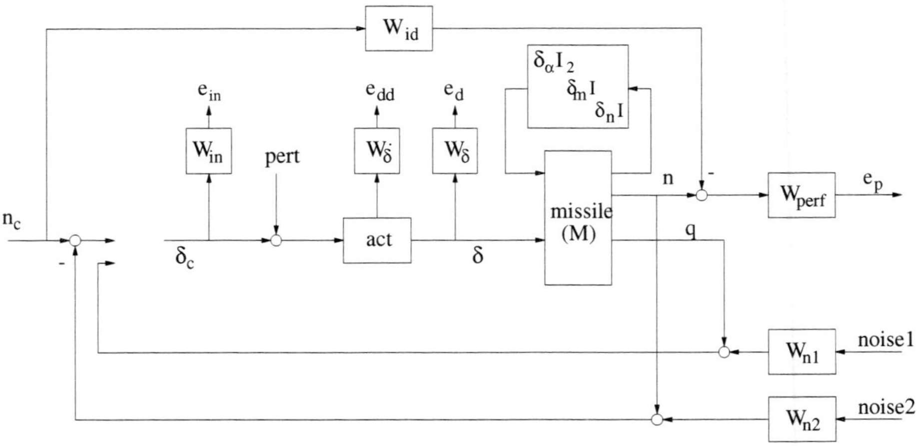


Figure 5.2: The controller synthesis interconnection structure

The control architecture for the missile problem is depicted in Figure 5.2. The tracking specification has been translated into an ideal acceleration model which the closed loop system should match. The ideal model comes from [Wu95] and is $W_{\text{id}}(s) = \frac{144(-0.05s+1)}{s^2+19.2s+144}$ for which the allowable error is weighted with $W_{\text{perf}}(s) =$

$\frac{0.5s+17.321}{s+0.0577}$. The low frequency gain of $W_{\text{perf}}$ is 300 to bound the tracking error to 0.33 %. To reflect the tail-fin deflection and deflection rate limits of 20 degrees and 25 degrees/s per $g$ respectively, constant weights $W_\delta = \frac{1}{20}$ and $W_{\dot\delta} = \frac{1}{25}$ have been chosen accordingly. Finally, noise weights $W_{\text{n1}} = 0.001$, and $W_{\text{n2}} = 0.001$ are used to reflect the measurement imperfections in pitch rate and normal acceleration.

## 5.4 Results

### 5.4.1 Design of the LPV Controller

In this section we will use the synthesis LMIs in equation (5.6) as derived previously. Solving the LMIs is done via basis functions and gridding of the parameter space. For the missile control problem, the Mach number $M$ is the remaining parameter for scheduling. The angle-of-attack $\alpha$, the aerodynamic coefficients $C_n$ and $C_m$ and the actuator were all assumed uncertain. Since the Mach number $M$ will be chosen to decrease linearly during the maneuver, we take a basis function as $f_1(M) = M$ (see also [AA96]). For the function $X(M)$, e.g., we thus have

$$X(M) = X_0 + X_1 M.$$

The other functions depending on the parameter $M$ are $Y(M)$, $Q(M)$, $K(M)$, $L(M)$, $M(M)$, $N(M)$, and have the same structure as $X(M)$.
The grid of the parameter set $M \in [2,\ 4]$ consisted of five points equally spaced between Mach $= 2$ and Mach $= 4$. As the Mach number will decrease from $M = 4$ to $M = 2$ in five seconds in the nonlinear simulations, the parameter rate was taken $|\dot M| < 0.5/s$. Further, we use a block diagonal scaling matrix $Q = \operatorname{diag}(Q_0, \dots, Q_4)$ arising from the uncertainty and the performance channels: the matrix $Q_1$ of dimension $2 \times 2$ for the uncertain $\alpha$, the two scalar blocks $Q_2$, $Q_3$ for the uncertainty in $C_n$, $C_m$, the scalar block $Q_4$ for the dynamic actuator uncertainty, and the $3 \times 3$ block $Q_0$ for the performance specification. In the first iteration the scalings are set to unity. Once convergence is achieved, the LMIs in equation (5.6) are solved for the last scaling $Q$ that was found in the iteration.
In Section 5.2.3 we revealed that choosing either $X$ or $Y$ constant leads to a controller that does not require a measurement of the parameter rate. For implementation in the missile control problem, both options were tested. The scheme using $X$ constant and $Y$ parameter dependent converged faster than the one where $X$ was parameter dependent and $Y$ constant. In 14 steps, a $\gamma$-value of 2.50 was reached for the first scheme in contrast to the other option where only value of 3.87 was reached after 20 iteration steps. In Figure 5.3 the achieved $\gamma$-value is set out against the number of iteration steps for both schemes.
The final controller synthesis was carried out on the choice $(X_0, Y(M))$ where $\gamma = 2.5$ was achieved. The scaling matrix $Q$ of the last iteration is used to synthesize the controller using the full synthesis LMIs. The problem size with

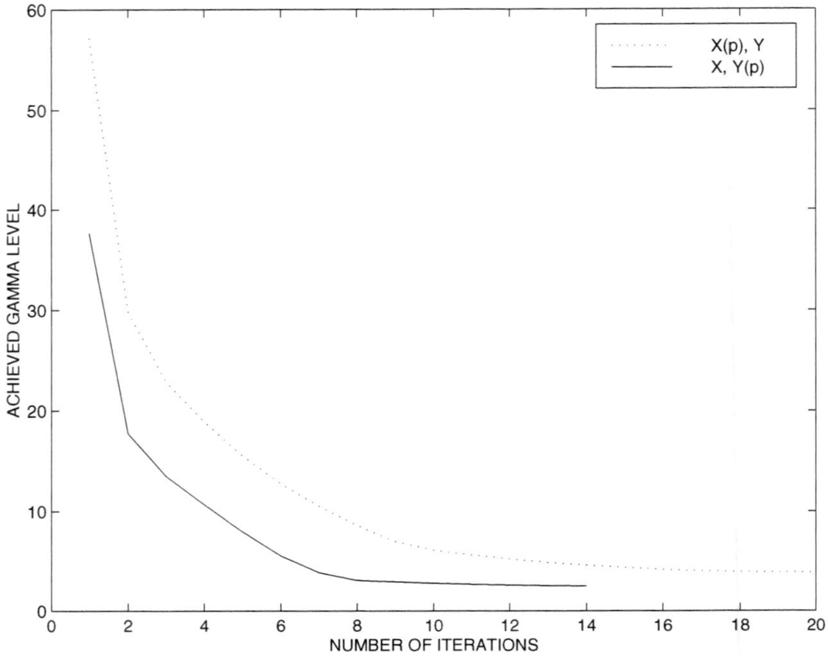


Figure 5.3: Achieved performance $\gamma(X, Y(M))_{14} = 2.50$ and $\gamma(X(M), Y)_{20} = 3.87$ against number of iterations

projected LMI's has 109 decision variables in contrast to the full LMI problem with 289 decision variables. The complete design cycle takes about 6 to 8 hours on a 200 MHz Pentium-pro machine.

The achieved $\gamma$-value was 2.51 and the test on a denser parameter grid (with twice the density) gave $\gamma$ values between 2.22 and 2.45. The achieved performance level given by the proposed scheme is comparable to results given in [Wu95] ($\gamma = 3.13$) and [Hel95] ($\gamma = 3.855$) with the difference that in our design parameter rate boundness and uncertainties are taken into account. Viewing the angle-of-attack as uncertain increases the achievable $\gamma$ level. Nevertheless, the introduced conservatism is compensated by taking into account the parameter rate $\dot{M}$. We end up with a lower $\gamma$-value and a controller implementation with less sensor information than given in [Wu95, Hel95].

Next, we illustrate how the missile problem behaves using different design methods and assumptions on the control and uncertainty structures. Running the same algorithm leaving out the aerodynamic uncertainties provided in ten iterations an achievable performance of $\gamma = 2.38$ for the case $X$ constant and $Y$ parameter dependent. Two other designs have been conducted using LFT gain scheduling approaches such as given in [BAG96, NSB96] using also the angle-of-attack $\alpha$ as a scheduling parameter. We achieved $\gamma$ values of 4.43

| Method | Schedule | Rate | $\Delta$ | $L_2$-gain |
|---|---|---|---|---|
| Helmersson [Hel95] LPV-LFT | $\alpha$, $M$ | $\infty$ | No | 3.85 |
| Partial-Scales [BAG96] LPV-LFT | $\alpha$, $M$ | $\infty$ | No | 4.43 |
| Full-Scales [NSB96] LPV-LFT | $\alpha$, $M$ | $\infty$ | No | 3.60 |
| Wu [Wu95] Grid | $\alpha$, $M$ | $\infty$ | No | 3.13 |
| This thesis Grid | $M$ | $0.5\,/s$ | $\alpha$, $C_m$, $C_n$, $Act$ | 2.51 |

Table 5.1: Comparision $L_2$-gain Different Designs

and 3.60 respectively. These LFT approaches assume the parameter rates of $\alpha$ and Mach unbounded. It is not surprising that this leads to more conservative $\gamma$-levels, which correspond to larger overshoots in the normal acceleration step command responses. Finally, to get an indication of the performance level that can be ideally obtained in each operating condition, robust $H_\infty$ controllers have been designed for frozen $\alpha$ and Mach operating points. The resulting $L_2$-induced norm ranged between $\gamma = 0.3$ in $(\alpha, M) = (20, 4)$ and $\gamma = 0.990$ in $(\alpha, M) = (0, 2)$.

We shall resume by comparing the various gain scheduling methods and results with the litterature [Pac94, Hel95, Wu95, BAG96, NSB96]. These are summarized in table (5.1).

### 5.4.2 Simulations

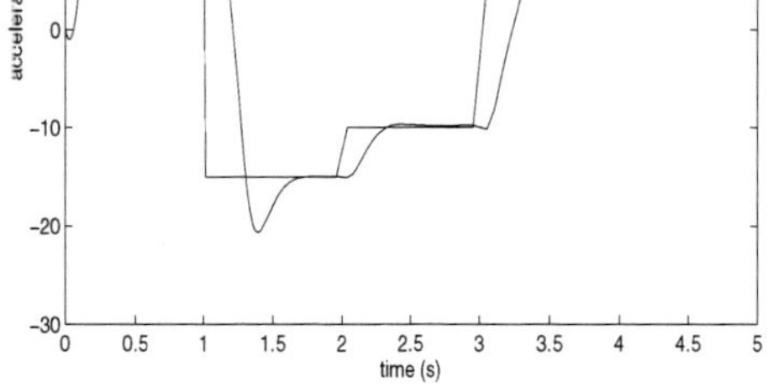


Figure 5.4: Normal acceleration $n_z$ for the commanded acceleration scenario $n_c$ of the LPV controlled missile.

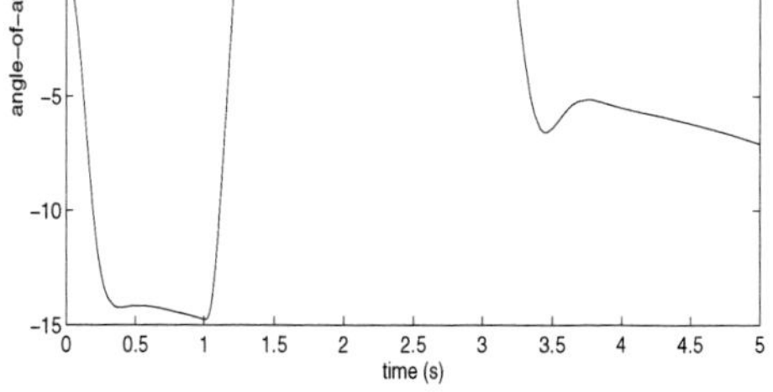


Figure 5.5: Angle-of-attack $\alpha$ for the commanded acceleration scenario $n_c$ of the LPV controlled missile.

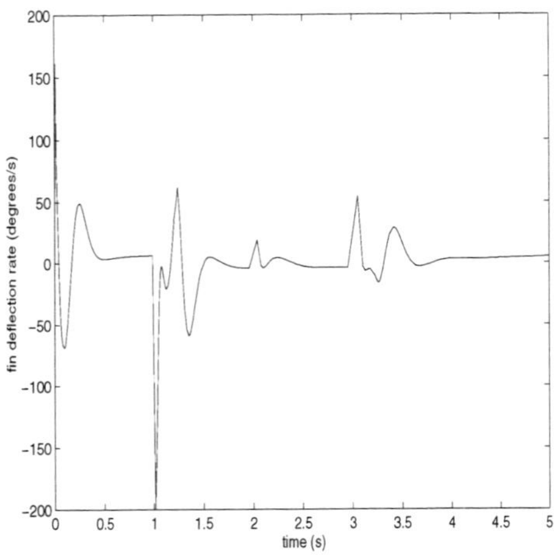


Figure 5.6: Fin deflection rate $\dot{\delta}$ for the commanded acceleration scenario $n_c$ of the LPV controlled missile.

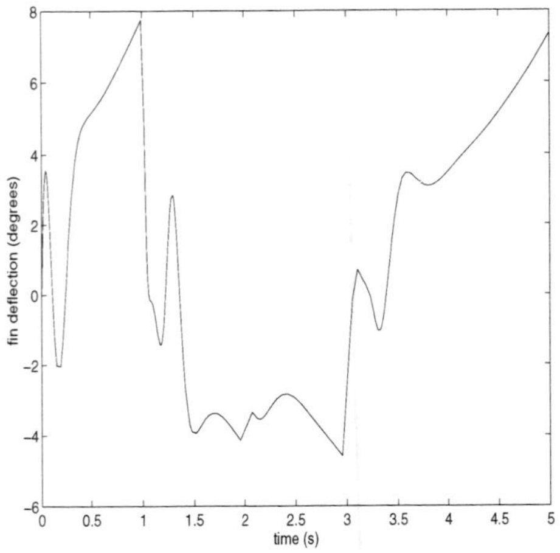


Figure 5.7: Fin deflection $\delta$ for the commanded acceleration scenario $n_c$ of the LPV controlled missile.

The nonlinear simulations of the LPV controlled missile are depicted in the Figures 5.4 to 5.8. The maneuver, during which the Mach number varies from 4 to 2 in 5 seconds, consists of a series of acceleration step commands as shown in Figure 5.4. The acceleration command response of the LPV controlled missile has a rise time that is less than the prescribed 0.35 s. The steady state error is within the required bounds. Overshoot characteristics are also within the limits. Only the step command from 30 g to $-15$ g causes a 3 % overshoot violation from the allowed 10 %. As a remedy one could try to redesign the weightings. A possible choice to enhance damping of the acceleration response is to increase the weight on the fin rate filter $W_{\dot{\delta}}$ since maximum fin rate is by far not reached in the nonlinear simulation. Also the performance filter could be adjusted to further penalize the overshoot (increase high frequency gain of the filter). However, we left the filters $W_{\text{id}}$ and $W_{\text{perf}}$ the same as in [Wu95] in order to be able to compare the results. The LPV controller synthesized here has a larger overshoot, but the responses are faster than the LPV design in [Wu95]. Moreover the LPV controller of [Wu95] shows angle-of-attack excursions beyond $-20$ degrees while also commanding four times more fin deflection rate during the $-45$ g command. Figures 5.5 and 5.6 show that the controller remains within the limits which might be interpreted as a rather conservative design in the sense that it doesn't use the available fin deflection 5.6 to enhance performance. It should be noted that the missile in [Wu95] runs along a slightly different Mach trajectory.

Finally, to demonstrate the robustness properties, Figure 5.8 shows the acceleration command responses for all combinations of the aerodynamic uncertainties. As can be seen from the figure, overshoot in the $-45$ g step is the most sensitive

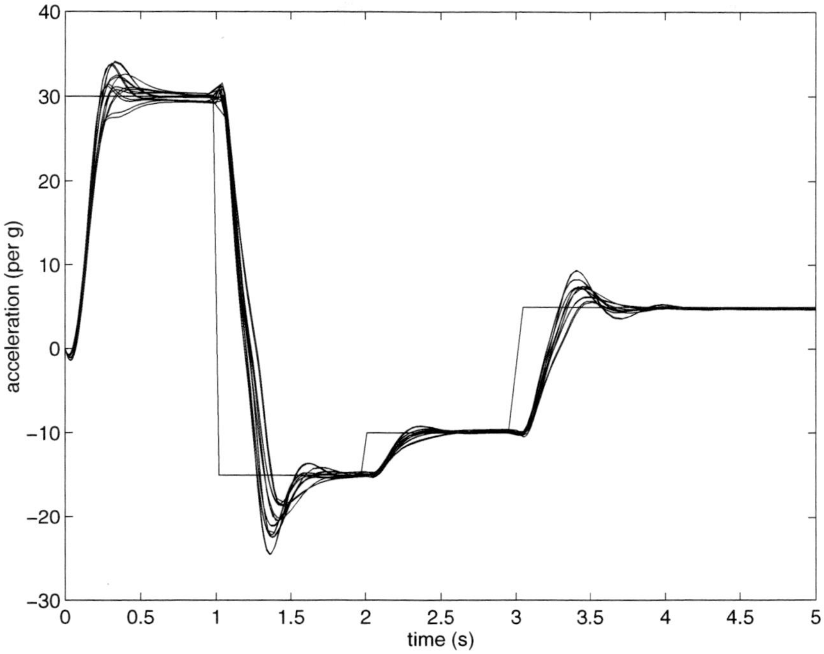


Figure 5.8: Command response for all combinations of perturbed aerodynamics

to uncertainties, while the other performance characteristics seem to behave well.

# 5.5 Conclusions

We have shown a controller synthesis technique for Linear Parameter Varying (LPV) that takes the boundedness of the parameter variation rates into account. This technique gives guaranteed stability and performance levels. Moreover, robustness against uncertainties has been incorporated via the use of scalings. Using basis functions and gridding, the synthesis problem is reduced to an iteration process of solving finitely many linear matrix inequalities. The method has been applied and tested on a missile benchmark problem. The nonlinear simulations have shown that the proposed method is successful in achieving the desired performance and robustness goals. Further research should be directed towards finding systematic procedures to choose the basis functions in the presented approach.

# Chapter 6

# Robust LPV Control with Unbounded Parameter Rates: Citation LPV-Design

*A flight control system design for a small business jet is presented. Robust linear parameter varying control is applied to the Delft laboratory aircraft, providing satisfactory Gibson handling quality requirements over a large part of the operating range.*

## 6.1 Introduction

Current engineering practice when designing flight control systems starts with many linear designs at fixed operating points. To meet the required performance and robustness levels over the entire operating range, the separately synthesized linear designs are interpolated along physically measurable parameters. In practice, stability and performance of the resulting gain scheduled control system is analyzed using tools dedicated to linear time invariant systems and time domain simulations. Guaranteed performance, however, necessitates formal tools as developed in the last decade. The research effort devoted to the study of formal gain scheduling techniques such as Linear Parameter Varying (LPV) control led to LPV techniques equipped with analysis and synthesis results in terms of Linear Matrix Inequality (LMI) conditions that guarantee the performance for gain scheduled systems.

In this chapter we use the LPV Linear Fractional Transformation (LFT) gain scheduling approach [LS97, SBB96] in combination with an intermediate $\mu$-analysis [BDG+93b] step to take stability robustness requirements into account.

The formal design procedure is shortly discussed in Section 6.2. Section 6.3 presents the actual design model together with the design objectives. It is shown how Level 1 handling quality requirements as proposed by the US Military Specifications [MIL87] and Gibson [Gib99] can be efficiently incorporated in the design. The obtained synthesis results are presented in Section 6.4. The analysis reveals to what extent the handling quality requirements have been met in terms of nominal performance (NP), robust stability (RS) and robust performance (RS).

## 6.2 Robust LPV-LFT Control

The general system interconnection for robust LPV-LFT controller synthesis is depicted in figure 6.1.

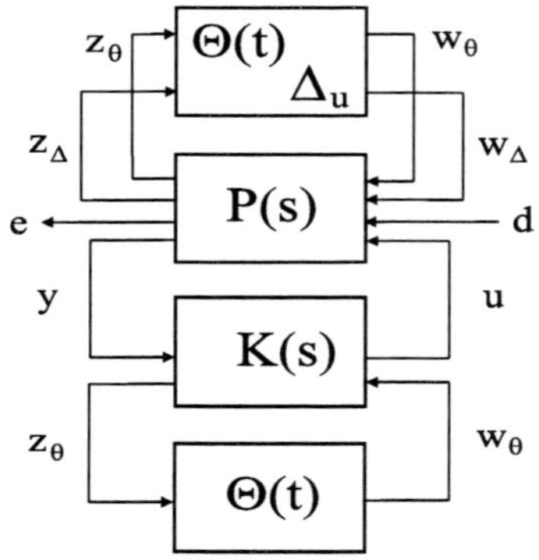


Figure 6.1: Uncertain Linear Parameter Varying Plant

The uncertain LPV-LFT system we want to control consists of an LTI part $P(s)$ depending in a linear fractional manner on the time-varying operator $\Theta$ (which is on-line measurable) and a norm bounded operator $\Delta_u$ representing a linear time invariant uncertainty. The LPV and uncertainty channels are appropriately dimensioned and related to each other by $z_\Delta = \Delta_u w_\Delta$ and $z_\theta = \Theta w_\theta$. The channel from $d$ to $e$ is called the performance channel and from $u$ to $y$ the control channel.

The characterization of an uncertain LPV-LFT system is as follows:

$$\begin{bmatrix} \dot{x} \\ z_\theta \\ z_\Delta \\ e \\ y \end{bmatrix} = \begin{bmatrix} A & B_\theta & B_\Delta & B_1 & B_2 \\ C_\theta & D_{\theta\theta} & D_{\theta\Delta} & D_{\theta 1} & D_{\theta 2} \\ C_\Delta & D_{\Delta\theta} & D_{\Delta\Delta} & D_{\Delta 1} & D_{\Delta 2} \\ C_1 & D_{1\theta} & D_{1\Delta} & D_{11} & D_{12} \\ C_2 & D_{2\theta} & D_{2\Delta} & D_{21} & D_{22} \end{bmatrix} \begin{bmatrix} x \\ w_\theta \\ w_\Delta \\ d \\ u \end{bmatrix} \tag{6.1}$$

$$
\begin{aligned}
w_\theta &= \Theta(t) z_\theta, \quad ||\Theta(t)|| \leq 1 \\
w_\Delta &= \Delta_u z_\Delta, \quad ||\Delta_u|| \leq 1
\end{aligned}
$$

The objective is to find a gain-scheduled controller $K(\Theta)$ with same LFT dependence in $\Theta(t)$ as the plant $P(s)$, given by:

$$
\begin{bmatrix} \dot{x} \\ u \\ z_\theta \end{bmatrix} = \begin{bmatrix} A_k & B_k & B_\theta \\ C_k & D_k & D_{k\,\theta} \\ C_\theta & D_{\theta\,k} & D_{\theta\theta} \end{bmatrix} \begin{bmatrix} x \\ u \\ w_\theta \end{bmatrix} \tag{6.2}
$$

$$
w_\theta = \Theta z_\theta
$$

The controller should stabilize the closed loop system depicted in figure 6.1 for the uncertainty block $\Delta_u$ and all admissible trajectories of $\Theta(t)$, while minimizing the $L_2$-induced norm of the performance channel.

To reduce conservatism in the nominal LPV problem, a set of constant scalings $L_\theta$ is introduced such that $L_\theta = \{L > 0 : L\Theta = \Theta L, \text{ for all admissible } \Theta\}$. Sufficient solvability conditions for the controller synthesis problem are expressed in terms of the scaled Bounded Real Lemma (BRL) [BAG96]. The nominal gain-scheduled control problem is solvable if there exist pairs of symmetric matrices $(X, Y)$, $[L_3, J_3]$ in $L_\theta$ such that the following Linear Matrix Inequalities hold:

$$
N_X^T \begin{bmatrix} AX + XA^T & XC_\theta^T & XC_1^T & B_\theta J_3 & B_1 \\ C_\theta X & -J_3 & 0 & D_{\theta\theta} J_3 & D_{\theta 1} \\ C_1 X & 0 & -I & D_{1\theta} J_3 & D_{11} \\ J_3 B_\theta^T & J_3 D_{\theta\theta}^T & J_3 D_{1\theta}^T & -J_3 & 0 \\ B_1^T & D_{\theta 1}^T & D_{11}^T & 0 & -I \end{bmatrix} N_X < 0 \tag{6.3}
$$

$$
N_Y^T \begin{bmatrix} A^T Y + YA & YB_\theta & YB_1 & C_\theta^T L_3 & C_1^T \\ B_\theta^T Y & -L_3 & 0 & D_{\theta\theta}^T L_3 & D_{1\theta} \\ B_1^T Y & 0 & -I & D_{\theta 1}^T L_3 & D_{11}^T \\ L_3 C_\theta & L_3 D_{\theta\theta} & L_3 D_{\theta 1} & -L_3 & 0 \\ C_1 & D_{1\theta} & D_{11} & 0 & -I \end{bmatrix} N_Y < 0 \tag{6.4}
$$

$$
\begin{bmatrix} X & I \\ I & Y \end{bmatrix} \geq 0 \tag{6.5}
$$

$$
\begin{bmatrix} J_3 & I \\ I & L_3 \end{bmatrix} \geq 0 \tag{6.6}
$$

$N_X$ and $N_Y$ are bases of the nullspaces of $\left[B_2^T\ D_{\theta 2}^T\ D_{12}^T\ 0\right]$ and $\left[C_2\ D_{2\theta}\ D_{21}\ 0\right]$, respectively.

In the problem formulation the rate of the parameter variation, $\dot{\Theta}(t)$, is assumed to be unbounded. This assumption introduces conservatism since the rate of variation is actually bounded [AA96, BWS98, BAG96].

The robustness issue in the LPV-LFT problem is taken into account according to the procedure outlined in [SBB96]. An improved performance level of the robust LPV controller is obtained via appropriate scaling of the time invariant uncertainty channels, using a separate $\mu$-analysis step. This scaling does not ruin the convexity of the LMI conditions given by 6.3. The procedure is referred to as D-K-D iteration. It does not provide a guaranteed optimal control solution, but it has proven to be of practical importance in an engineering context.

1. Use $\mu$-synthesis to find a controller $K(s)$ and a scaling $D$ for the scaling and initialize the scaling to identity. This leads to the normalized scaling $L_\Delta$.

2. Form an *inner* $D$ scaling, $D_{in}$,

$$D_{in} = \begin{bmatrix} I & 0 \\ 0 & L_\Delta \end{bmatrix}$$

and the system

$$T_{in}(s) = D_{in} F_l \left[P_a(s), K(s)\right] D_{in}^{-1}$$

$P_a(s)$ is defined as

$$P_a(s) = \begin{bmatrix} d_{11} & d_{12} \\ d_{21} & 0 \end{bmatrix} + \begin{bmatrix} c_1 \\ c_2 \end{bmatrix} [sI - A]^{-1} \begin{bmatrix} b_1^T \\ b_2^T \end{bmatrix}^T \tag{6.7}$$

with

$$b_1 = \begin{bmatrix} 0 & 0 & B_\theta & B_1 \end{bmatrix}, \quad b_2 = \begin{bmatrix} B_2 & 0 \end{bmatrix} \quad c_1 = \begin{bmatrix} 0 \\ C_\theta \\ C_1 \end{bmatrix}$$

$$d_{11} = \begin{bmatrix} 0 & 0 & 0 \\ 0 & D_{\theta\theta} & D_{\theta 1} \\ 0 & D_{1\theta} & D_{11} \end{bmatrix}, \quad d_{12} = \begin{bmatrix} 0 & I \\ D_{\theta 2} & 0 \\ D_{12} & 0 \end{bmatrix}$$

$$c_2 = \begin{bmatrix} C_2 \\ 0 \end{bmatrix}, \quad d_{21} = \begin{bmatrix} 0 & D_{2\theta} & D_{21} \\ I & 0 & 0 \end{bmatrix}$$

3. Use LMIs (6.3)-(6.6) to find an *outer* $D$ scaling, $D_{out}$,

$$D_{out} = \begin{bmatrix} (J_3)^{\frac{1}{2}} & 0 \\ 0 & I \end{bmatrix}$$

such that

$$\gamma = \left\| D_{out} T_{in} D_{out}^{-1} \right\|_\infty$$

is minimized.

4. Design an $H_\infty$ controller $K(s)$ for the system

$$\begin{aligned} P_s(s) &= \begin{bmatrix} F d_{11} F^{-1} & F d_{12} \\ d_{21} F^{-1} & d_{22} \end{bmatrix} + \\ & \begin{bmatrix} F c_1 \\ c_2 \end{bmatrix} [sI - A]^{-1} \begin{bmatrix} F^{-T} b_1^T \\ b_2^T \end{bmatrix}^T \end{aligned}$$

with $F$=Choleski decomposition of the scaling $J_3$.

5. Form the outer system $T_{out}(s) = D_{out} F_l \left[P_a(s), K(s)\right] D_{out}^{-1}$ and find over a frequency range scalars $m_\theta$ and structured matrices $M_\Delta$ minimizing

$$\mu = \left\| \begin{bmatrix} m_\theta I & 0 \\ 0 & M_\Delta \end{bmatrix} T_{out} \begin{bmatrix} m_\theta I & 0 \\ 0 & M_\Delta \end{bmatrix}^{-1} \right\|_\infty$$

6. Normalize $M_\Delta$ at each frequency by $m_\theta$ and obtain $L_\Delta$ by fitting a transfer function to the normalized $M_\Delta$'s.

7. Form with this new $L_\Delta$ a new $D_{in}$ and $T_{in}$ and return to step 3. The procedure is ended when $\mu$ has reached its minimum.

## 6.3 Design Objectives

The aircraft model is a second order linear parametrically varying model of the longitudinal short period dynamics of a Cessna Citation 500:

$$\begin{bmatrix} \dot{q} \\ \dot{\alpha} \\ \hline q \\ \alpha \end{bmatrix} = \left[\begin{array}{cc|c} M_q(V,h) & M_\alpha(V,h) & M_{\delta_e}(V,h) \\ Z_q(V,h) & Z_\alpha(V,h) & Z_{\delta_e}(V,h) \\ \hline 1 & 0 & 0 \\ 0 & 1 & 0 \end{array}\right] \begin{bmatrix} q \\ \alpha \\ \hline \delta_e \end{bmatrix} \tag{6.8}$$

with $\alpha$ $[rad]$ the angle of attack, $q$ $[rad/sec]$ the pitch rate and $\delta_e$ $[rad]$ the elevator deflection. It is assumed that all aircraft states are available as measurements. The measurements reflecting the LPV dependence in the aircraft model are airspeed $V$ $[m/s]$ and altitude $h$ $[ft]$. The parameter dependent state

space coefficients in equation 6.8 are obtained through polynomial interpolation of the linearizations over the parameter space $V \in [100m/s\ 175m/s] \times h \in [5000ft\ 30000ft]$. For controller synthesis purposes the parameter range is normalized to $\delta_V \in [-1\ 1]$ and $\delta_h \in [-1\ 1]$.
The resulting system depends on the parameters $\delta_V$ and $\delta_h$. The functions representing the state space coefficients' dependence on $\delta_V$ and $\delta_h$ are:

$$\begin{aligned}
M_q = & -3.29 * 10^{-3}\ \delta_V^2\ \delta_h^2 + 2.79 * 10^{-4}\ \delta_V^2\ \delta_h + 9.61 * 10^{-3}\ \delta_V^2 \\
& - 4.07 * 10^{-2}\ \delta_V\ \delta_h^2 + 2.58 * 10^{-1}\ \delta_V\ \delta_h - 6.46 * 10^{-1}\ \delta_V \\
& - 1.49 * 10^{-1}\ \delta_h^2 + 9.51 * 10^{-1}\ \delta_h - 2.19 \\
Z_q = & 2.39 * 10^{-6}\ \delta_V^2\ \delta_h^2 + 6.12 * 10^{-5}\ \delta_V^2\ \delta_h + 2.63 * 10^{-4}\ \delta_V^2 \\
& + 2.39 * 10^{-6}\ \delta_V\ \delta_h^2 - 1.53 * 10^{-5}\ \delta_V\ \delta_h - 7.01 * 10^{-4}\ \delta_V \\
& - 1.32 * 10^{-3}\ \delta_h^2 + 8.25 * 10^{-3}\ \delta_h + 9.81 * 10^{-1} \\
M_\alpha = & 7.56 * 10^{-1}\ \delta_V^2\ \delta_h^2 + 7.61 * 10^{-1}\ \delta_V^2\ \delta_h - 1.78\ \delta_V^2 \\
& - 8.05 * 10^{-1}\ \delta_V\ \delta_h^2 + 3.73\ \delta_V\ \delta_h - 7.27\ \delta_V - 1.41\ \delta_h^2 \\
& + 5.71\ \delta_h - 1.17 * 10^1 \\
Z_\alpha = & -3.12 * 10^{-2}\ \delta_V^2\ \delta_h^2 - 6.06 * 10^{-2}\ \delta_V^2\ \delta_h + 8.58 * 10^{-2}\ \delta_V^2 \\
& - 3.91 * 10^{-2}\ \delta_V\ \delta_h^2 + 9.26 * 10^{-2}\ \delta_V\ \delta_h - 3.26 * 10^{-1}\ \delta_V \\
& - 8.00 * 10^{-2}\ \delta_h^2 + 5.51 * 10^{-1}\ \delta_h - 1.31 \\
M_{\delta_e} = & -4.68 * 10^{-1}\ \delta_V^2\ \delta_h^2 + 5.23 * 10^{-1}\ \delta_V^2\ \delta_h - 4.39 * 10^{-1}\ \delta_V^2 \\
& - 5.60 * 10^{-1}\ \delta_V\ \delta_h^2 + 2.98\ \delta_V\ \delta_h - 9.05\ \delta_V \\
& - 7.19 * 10^{-1}\ \delta_h^2 + 6.98\ \delta_h - 1.96 * 10^1 \\
Z_{\delta_e} = & 4.60 * 10^{-4}\ \delta_V^2\ \delta_h^2 + 3.00 * 10^{-4}\ \delta_V^2\ \delta_h - 7.20 * 10^{-5}\ \delta_V^2 \\
& - 2.71 * 10^{-3}\ \delta_V\ \delta_h^2 + 1.49 * 10^{-2}\ \delta_V\ \delta_h - 3.58 * 10^{-2}\ \delta_V \\
& - 8.79 * 10^{-3}\ \delta_h^2 + 5.51 * 10^{-2}\ \delta_h - 1.32 * 10^{-1}
\end{aligned}$$

The parameter dependent state space system 6.8 is then reformulated as a Linear Fractional Transformation (LFT) format, using the PUM toolbox for MATLAB [LTBS93, TL93]:

$$\begin{bmatrix} \dot{x} \\ y \\ z_\theta \end{bmatrix} = \begin{bmatrix} A & B & B_2 \\ C & D & D_{12} \\ C_2 & D_{21} & D_{22} \end{bmatrix} \begin{bmatrix} x \\ u \\ w_\theta \end{bmatrix} \tag{6.9}$$

$$w_\theta = \Theta z_\theta$$

where $\Theta = diag(\Delta_V, \Delta_h) \in (\delta_V I_6, \delta_h I_4)$.

The elevator position is changed via an actuator with first order dynamics and a bandwidth of 15 rad/s:

$$G_{act}(s) = \frac{15}{s + 15} \tag{6.10}$$

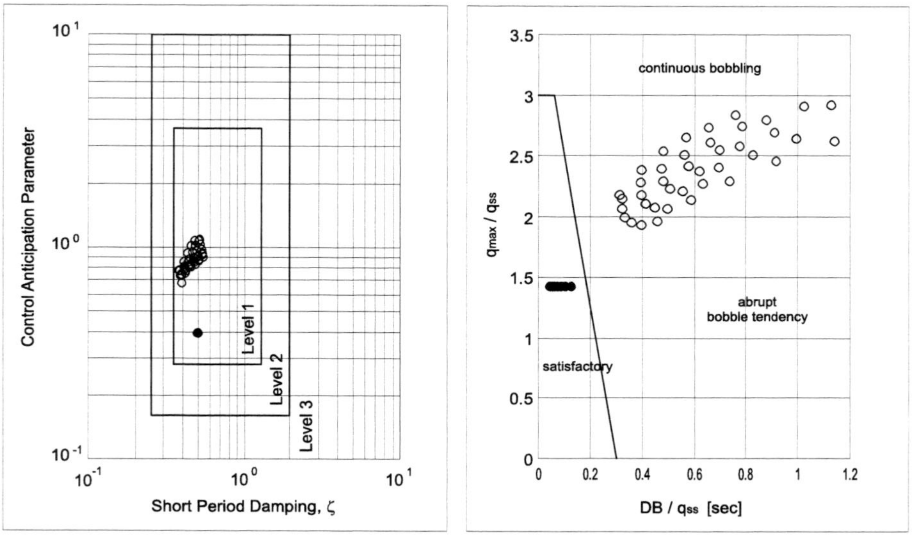


Figure 6.2: CAP and Dropback Criterion Parameters for the Actual Aircraft Model and the 'Ideal' Model

It is assumed that the actuator may give a position error of 5% up to a frequency of 10 rad/s, at higher frequencies this may be more. This is represented by the following transfer function:

$$W_{act}(s) = \frac{s + 10}{s + 200} \tag{6.11}$$

This perturbation also accounts for higher order unmodeled dynamics of the plant, like unknown structural modes. The weighting function is used to normalize the unknown perturbation $\delta$. At any frequency $\omega$ the magnitude of $W_{act}$ represents the relative uncertainty level in the actuator model.

The overall performance objective is to robustly track pitch rate commands with predicted Level 1 handling qualities [MIL87] without excessive control effort. Therefore we define a handling qualities model that we want to match. Level 1 handling qualities specifications for pitch rate responses in the time domain, defined in [LS97] are met (at fixed operating conditions) by the parameterized transfer function:

$$G_{id}(s) = \frac{\omega_{id}^2 (T_{id} s + 1)}{s^2 + 2\zeta_{id}\omega_{id} s + \omega_{id}^2} \tag{6.12}$$

where $\omega_{id} = \frac{3}{100}V$, $1/T_{id} = \frac{3}{4}\omega_{id}$ and $\zeta_{id} = 0.5$. The ideal short period frequency $\omega_{id}$ and short period numerator zero $1/T_{id}$ both depend on the airspeed $V$ while the damping $\zeta_{id}$ is constant.

For purpose of LPV control the parameter dependent handling qualities model $G_{id}(s)$ is transformed into an LFT model depending on the measurable normalized airspeed with block structure $\delta_V I_3$.

Recall that the pitch rate transfer function at fixed operating condition can be written as:

$$\frac{q(s)}{\delta_e(s)} = \frac{(M_\alpha Z_{\delta_e} - Z_\alpha M_{\delta_e})(\frac{M_{\delta_e}}{M_\alpha Z_{\delta_e} - Z_\alpha M_{\delta_e}} s + 1)}{s^2 - (Z_\alpha + M_q)s + M_q Z_\alpha - M_\alpha Z_q}$$

$$= \frac{K_\theta(sT_{\theta_2} + 1)}{s^2 + 2\zeta_{sp}\omega_{sp}s + \omega_{sp}^2} \tag{6.13}$$

with short period frequency and damping:

$$\omega_{sp} = \sqrt{M_q Z_\alpha - M_\alpha Z_q},\ \zeta_{sp} = \frac{-(Z_\alpha + M_q)}{2\omega_{sp}},$$

and pitch attitude gain and pitch rate numerator time constant:

$$K_\theta = M_\alpha Z_{\delta_e} - Z_\alpha M_{\delta_e} \text{ and } T_{\theta_2} = \frac{M_{\delta_e}}{K_\theta}$$

The pitch rate numerator time constant is often called the flight path time delay since it is the only parameter defining the lag between the flight path angle $\gamma$ and the attitude angle $\theta$ according to:

$$\frac{\gamma(s)}{\theta(s)} = \frac{1}{1 + T_{\theta_2}s}$$

With these equations the general short period parameters $\omega_{sp}$, $\zeta_{sp}$ and $T_{\theta_2}$ can be calculated for each operating condition. Notice that $T_{\theta_2}$ depends on the aerodynamic characteristics of the aircraft. It is predominantly dependent on the angle of attack lift curve slope since $T_{\theta_2} = -\frac{1}{Z_\alpha}$. Physically this parameter can be controlled by the application of appropriate direct lift control for any given flight condition. In control engineering terms the parameter $1/T_{\theta_2}$ is the transfer function zero of the elevator to pitch rate transfer function. Placing the system zeros can only be achieved by filtering and feedforward compensation which must be tuned for every flight condition to insure a satisfactory command response shape over the desired operating envelope. Robust and LPV control provide in this automatically by the choice of a two degrees of freedom control architecture consiting of a feedforward and feedback compensator.

Once these parameters are known the value of the Control Anticipation Parameter (CAP) can be calculated [HMA+82]:

$$CAP = \frac{\omega_{sp}^2}{\frac{V}{g}\frac{1}{T_{\theta_2}}} \tag{6.14}$$

CAP is a measure of the ability of a pilot to anticipate the medium term normal acceleration response from the initial pitch response and it has been shown that its value should be within certain limits to ensure adequate handling qualities.

Handling qualities can also be examined using the Dropback criterion [Gib99]. The Dropback criterion was introduced as one of the time domain design guidelines for highly augmented fighter aircraft. It is defined in terms of limiting values on the pitch rate overshoot ratio $[\frac{q_{max}}{q_{ss}}]$ and the ratio of attitude dropback to steady state pitch rate $[\frac{DB}{q_{ss}}]$. The definition of these parameters is shown in figure 6.3, which depicts typical responses of a rate command type aircraft to a step stick input. The parameters involved are the peak pitch rate $q_{max}$, and the attitude Dropback, DB, normalized to the steady state pitch rate, $q_{ss}$, as defined in figure 6.3. Dropback is computed as the difference between the peak pitch attitude $\theta_{max}$ at the time the stick is released and the steady state attitude $\theta_{ss}$ after the stick is released. A positive value of this difference is referred to as Dropback, while a negative value is named Overshoot.

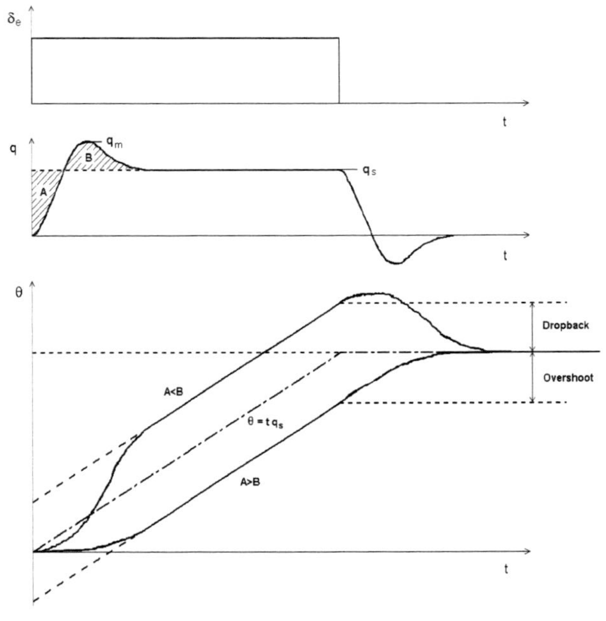


Figure 6.3: Definition of Gibson's Dropback Criterion

Regions of typical pilot comments are defined in the criterion plane $(DB/q_s, q_{max}/q_s)$, see figure 6.2. The criterion is mapped to qualitative response descriptions such as abruptness, sluggishness and bobbling. Negative Dropback indicates sluggishness, while large positive values of the Dropback indicate abrupt and bobbling tendencies. Values of Dropback and steady state pitch rate ratio between 0 and 0.3 sec and pitch rate overshoot ratio between 1 and 3 are usually considered good. The physical meaning of the criterion is that the satisfactory region is associated with predictability of the attitude response after a corrective pilot command. The amount of Dropback and Overshoot indicate how well the aircraft responds or 'follows' stick commands. The criterion was developed for fighters for accurate pitch tracking tasks. Ideally there should be no attitude dropback to stick commands, and the aircraft nose should exactly follow the stick. However, such a characteristic can lead to a sluggish flight path response which is not desirable for approach and landing tasks.

The time domain Dropback criterion defined in terms of limiting values on pitch rate overshoot ratio $[\frac{q_{max}}{q_{ss}}]$ and on the ratio of attitude dropback to steady state pitch rate $[\frac{DB}{q_{ss}}]$, which can be expressed in frequency domain short period parameters as:

$$DB/q_{ss} = T_{\theta_2} - 2\zeta_{sp}/\omega_{sp} \tag{6.15}$$

The handling qualities of the ideal handling qualities model meet Level 1 requirements of the CAP criterion and the Dropback criterion. This can be seen from figure 6.2, where the CAP and Dropback criterion values of the ideal handling qualities model are shown and calculated at various airspeeds according to:

$$CAP_{id}(V) = \frac{\omega_{id}^2}{\frac{V}{g}\frac{1}{T_{id}}} = 0.4 \text{ and } \zeta_{id} = 0.5 \tag{6.16}$$

$$(DB/q_{ss})_{id}(V) = T_{id} - 2\zeta_{sp}/\omega_{id} = \frac{100}{9V} \text{ and } [\frac{q_{max}}{q_{ss}}] = 1.4 \tag{6.17}$$

Notice that in the left hand side of figure 6.2 which represents the $(\zeta sp, CAP_{sp})$-space, the chosen parameter dependent ideal model $G_{id}(V)$ leads to one single handling quality point $(\zeta_{id}, CAP_{id}) = (0.5, 0.4)$.

In figure 6.2 CAP and Dropback criterion parameters are also plotted for the open loop aircraft model given by equation 6.8 for 42 flight conditions within the parameter space $V \in [100m/s\ 175m/s] \times h \in [5000ft\ 30000ft]$. For all operating conditions examined, the CAP value lies within the Level 1 region. Hence, according to the CAP criterion, the aircraft dynamics are predicted to have good handling qualities.

According to the Dropback criterion, for the selected operating conditions the aircraft does not have satisfactory handling qualities. All 42 evaluated flight cases in the right hand side of figure 6.2 show continuous bobbling tendency. This can be annoying for the pilot during tight tracking tasks and even might lead to PIO in the case nonlinearities such as rate limiters are included in the actuator dynamics. Notice further that the ideal system would have an ideal $K/s$-like pitch attitude response if the ideal damping term $\zeta_{id}$ would have been chosen to be equal to $2\omega_{id}T_{id}$. This condition can be directly obtain by constraining the ideal Dropback parameter to be zero at each flight condition. However this choice would imply a considerably high value of the ideal damping since $\zeta_{id} = 2\omega_{id}T_{id} = \frac{8}{3} \approx 2.66$.

With the open loop aircraft model and the ideal handling qualities model defined, we will set up the closed loop interconnection structure in which the controller synthesis will be performed. Figure 6.4 shows the controller synthesis interconnection structure used to synthesize a robusty stabilizing LPV controller $K(V, h)$.

Since we want to design a pitch rate controller we will feed the pitch rate state $q$ back to the controller. The commanded pitch rate $q_{cmd}$ is provided to the controller directly. The output of the controller will be the input to the actuator.

The next step is to define the design specifications and incorporate them in the interconnection structure. The design specifications are translated to mathematical objectives by defining weighting functions [MTB97].
To get a well conditioned problem, the input of the system - the pitch rate demand $q_d$ - is weighted with its maximum expected value, $\pm 5$ degrees/second. Hence the frequency independent input weighting is

$$W_{in} = 5/57.3 \tag{6.18}$$

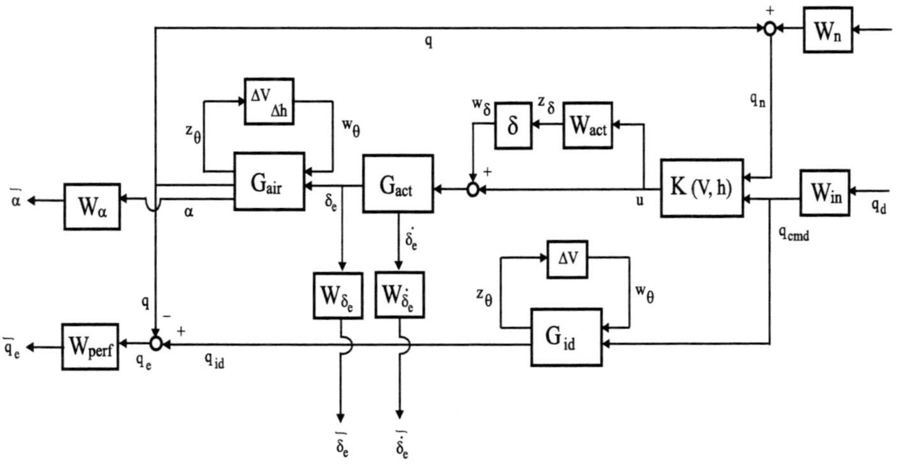


Figure 6.4: Closed Loop Interconnection Structure for Controller Synthesis

**Nominal Performance Objective: NP**

**Parameter Dependent Tracking Objective: NP 1**
The pitch rate tracking error between the parameter dependent ideal model response and the parameter dependent closed loop system response $q_e = q_{id} - q$, defines the actual nominal parameter dependent performance of the closed loop system. To emphasize how large and up to what frequency the error reduction on $q_e$ should last a frequency dependent weighting function $W_{perf}(s)$, reflecting the frequency content of the tracking objective, is placed on the error signal $q_e$. At low frequencies this weighting function need to have high gain in order to keep the steady state value of the error $q_e$ small. Beyond the bandwidth $\omega_B > \omega_{id}$ of the ideal model equation 6.12 the tracking error weight is released and the magnitude of $W_{perf}(s)$ rolls off. The weighting function $W_{perf}(s)$ is defined as a first order model [SP96]:

$$W_{perf}(s) = \frac{\frac{1}{M_p}s + \omega_B}{s + \omega_B A_{ss}} \tag{6.19}$$

For our design the parameters $(M_p, \omega_B, A_{ss})$ is chosen as (1.5, 10, 0.01). This means that a steady state tracking error of 1 % is allowed with respect to the normalized commanded pitch rate input $q_d$. The weighted tracking error is compensated for the reference input weighting. The resulting tracking error weighting function then becomes:

$$W_{perf}(s) = \frac{57.3}{5}\frac{\frac{1}{1.5}s + 10}{s + 0.1} \tag{6.20}$$

This requirement should be achieved uniformingly for all flight conditions within the parameter space $V \in [100m/s\ 175m/s] \times h \in [5000ft\ 30000ft]$.

**Control Effort Minimization: NP 2**
Tracking should not be achieved at the cost of too high control effort. Therefore both, the elevator deflection and the elevator deflection rate are penalized. The deflection and rate are weighted with the inverse of their maximum allowed values:

$$\begin{aligned} W_{\delta_e} &= 1/\delta_{e_{max}} = 57.3/15\ rad^{-1} \\ W_{\dot{\delta}_e} &= 1/\dot{\delta}_{e_{max}} = 57.3/30\ (rad/s)^{-1} \end{aligned} \tag{6.21}$$

**Angle of Attack Limitation: NP 3**
To prevent the closed loop system from stalling during a full pitch rate command it is necessary to provide some angle of attack limiting function. As in the previous weighting function choice we limit the angle of attack $\alpha$ with the inverse of its maximum allowable value:

$$W_\alpha = 1/\alpha_{max} = 57.3/20\ rad^{-1} \tag{6.22}$$

**Measurement Noise Rejection: NP 4**
Finally, a noise filter $W_n(s)$ is defined to scale the normalized measurement noise $n$ as a function of the frequency:

$$W_n = 1 * 10^{-6} \frac{100s + 1}{5s + 1} \tag{6.23}$$

**Robust Stability Objective: RS**

The parameter dependent controller $K(V, h)$ should robusty stabilize the LPV system shown in figure 6.4 with repsect to the input uncertainty on the actuator system.

**Robust Performance Objective: RP**

The robust performance objective is to achieve over the entire uncertainty set on the actuator dynamics robust stabilization (RS), while simultaneously all parameter dependent nominal performance objectives (NP) are met.

## 6.4 LPV Controller Synthesis

The controller synthesis interconnection structure shown in figure 6.4 will be used to synthesize a robust LPV controller by executing the D-K-D iteration. Before the LPV controller is synthesized, first an indication of the achievable LTI robust performance level need to be determined through proper choice of the weighting functions. This can ideally be achieved for fixed operating conditions by performing individual $\mu$-synthesis designs for all 42 flight conditions within the parameter space $V \in [100m/s\ 175m/s] \times h \in [5000ft\ 30000ft]$.
By freezing the time-varying parameters $V$ and $h$ at fixed values within the operating envelope (indicating zero rate of variation of the parameter) we get a set linear time invariant (LTI) uncertain systems.

For the 42 frozen LTI systems we design robust $H_\infty$ controllers applying $\mu$-synthesis. The performance and uncertainty weighting functions are tuned until the associated closed-loop system nominal performance (NP), robust stability (RS) and robust performance (RP) are assessed [BDG⁺93b]. Figure 6.5 shows the achieved levels of NP,RS and RP. Since these levels lie well below 1, NP, RS and RP are achieved for every operating condition.

At this point the robust LPV controller $K(V, h)$ can be designed, using the synthesis procedure described previously. The controller uses for scheduling in total 13 channels, 9 for speed and 4 for the altitude. As earlier stated, the

LPV controller is designed assuming an unbounded rate of variation. The scaling procedure for the parameter dependent and uncertainty channels should be stopped when no further reduction of $\mu$ can be obtained. In our design case the best achievable value of $\mu$ is reached after two iterations. Since the LMIs turn out to be solvable, the designed LPV controller $K(V, h)$ achieves robust stability over the entire flight envelope. Nominal performance is not achieved. This means that there exist arbitrarily fast parameter variations that can deteriorate the performance level with a factor 2. Robust performance however, is not achieved, since the best $L_2$ induced norm performance is 2.69.

In order to analyze the levels of NP, RS and RP that are achieved for fixed operating conditions, we close the scheduling channels of both the aircraft and ideal model and the controller with the same parameter vector $\Theta = diag(\delta_V I_9, \delta_h I_4)$. Since this results in an LTI system, we can test the achieved levels of NP, RS and RP. The results of the analysis for each operating condition are shown in figure 6.6. Two important observations can be made from this analysis:

- When we compare the levels of NP, RS and RP of the LPV controller at fixed operating conditions with those of the LTI controllers, we find that the LTI controllers achieve better performance levels. This indicates that taking into account the parameter dependence of the system when designing the controller leads to performance degradation. This inevitable trade-off between robustness and performance is characteristic for robust LPV control since the performance objective must now be achieved globally for all flight conditions within the parameter space $V \in [100m/s\ 175m/s] \times h \in [5000ft\ 30000ft]$.

- From figure 6.6 it can be seen that the robust performance level achieved by the LPV controller at fixed operating conditions lies far beneath the best $L_2$ induced norm performance, which is 2.69. Apparently the unrealistic assumption of unbounded rate on the parameter variation also causes a performance degradation. This means that the performance of the LPV controller can be improved by weakening the assumption of unbounded rate on the parameter variation. This could be done by either dynamic filters on parameter measurements which would increase the state dimension of the controller or by using the much more computationally involving LPV method with parameter dependent Lyapunov functions [AA96, BWS98].

The performance of the LPV controller is further investigated by applying the Dropback criterion [Gib99] to the system without actuator uncertainty, at fixed flight conditions. The calculated values of $[\frac{q_{max}}{q_{ss}}]$ and $[\frac{DB}{q_{ss}}]$ are shown in figure 6.7. From this figure it can be seen that, according to the Dropback criterion, the system has satisfactory handling qualities for airspeeds up to 137.5 m/s. This indicates that even if the system does not achieve robust performance (i.e. adequate tracking of the ideal model) the system may have good handling qualities.

## 6.5 Conclusion

It has been shown that the LPV and the $\mu$-design framework are well suited for the efficient synthesis of gain-scheduled controllers with desired robustness and parameter varying handling quality properties. The LPV design activity by itself can be seen as a formal continuation of the robust LTI design process. For a meaningful LPV design it matters how the parameter dependence of the ideal handling qualities have to be modelled. We have chosen to simultaneously achieve LPV CAP and Dropback requirements by a suitable choice of the parameter dependent ideal model. According to these criteria the LPV design method provides in one gain scheduled controller $K(V, h)$ stabilizing the plant globally. The plant achieves robustly the time varying stability criterion dictated by the Scaled Bounded Real Lemma. This means that the obtained controller has inherited stability guarantees for arbitrarily fast parameter changes which is a demanding constraint. By bounding the paramater rate to physical realistic levels it seems that the desired performance objectives can be met globally, since the design for zero parameter rate achieves robust performance globally.

It is necessary in a future work to direct the research towards an analysis of the system including nonlinear, uncertain and time-varying elements in a setting based on Integral Quadratic Constraints (IQCs) [MR97]. This will allow to derive stability conditions of the designed LPV system in the face of a nonlinear pilot model, an actuator model with rate limiters and a controller with uncertain scheduling parameters.

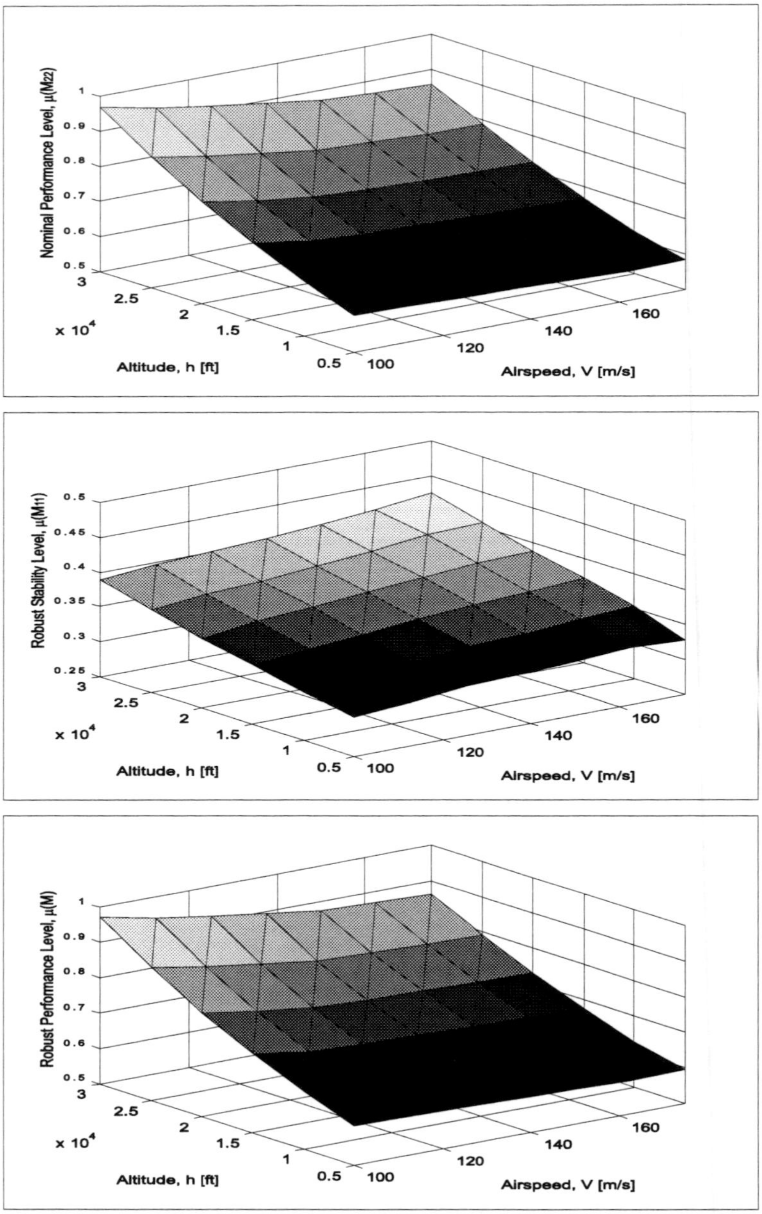


Figure 6.5: Achieved levels of NP,RS, and RP for fixed operating conditions (LTI controllers)

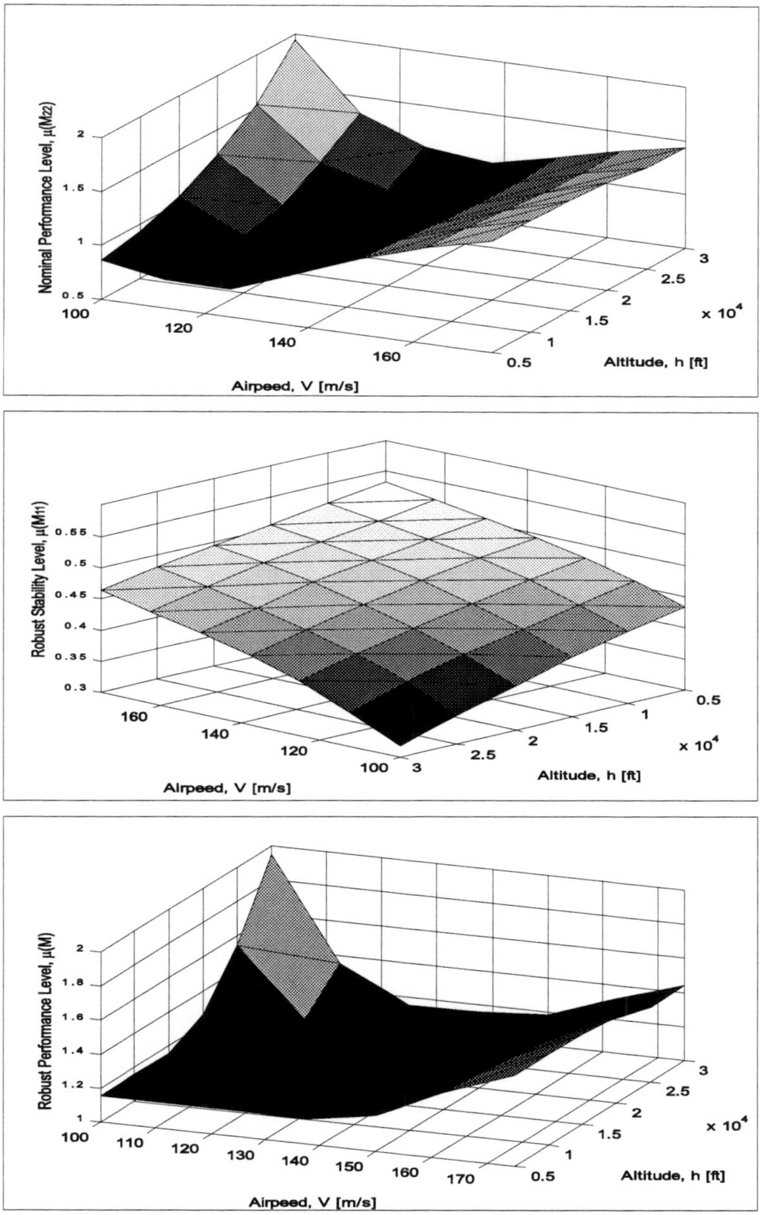


Figure 6.6: Achieved levels of NP,RS, and RP for fixed flight conditions (LPV controller)

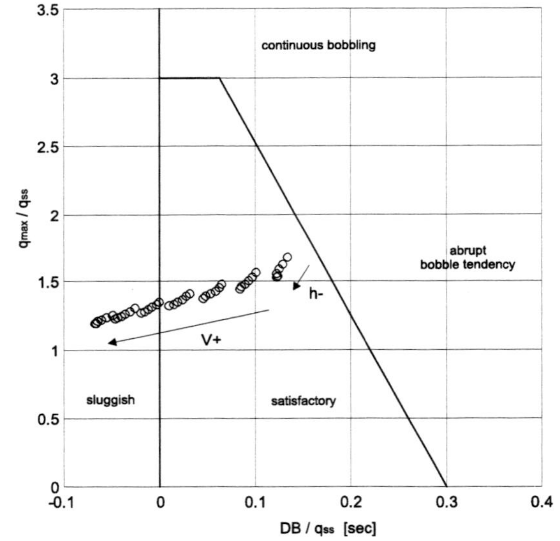


Figure 6.7: Dropback Criterion for System with LPV Controller

# Chapter 7

# Conclusions & Recommendations

## 7.1 Review of Results of the Various Designs

The object of the present thesis was to assess the possibilities of applying $\mu$-synthesis and various robust LPV techniques to realistic flight control problems. Questions to be answered concerning both the design process and the quality of the final design were

- How flexible is the designer in the choice of the system architecture
- How flexible is the designer in the inclusion of specific aeronautical design specifications
- How good is the design solution
- How flexible is the method with respect to redesign in the face of evolving design requirements.

The design problem is characterized by a number of fundamental trade-offs between conflicting design specifications. It was shown how to incorporate, using $\mu$-synthesis for LTI systems and Robust LPV synthesis for parameter dependent systems how to incorporate the following objectives for the synthesis of flight control systems:

- **Performance**
- **Robustness**
- **Robust performance**

Before drawing some general conclusions, we summarize the findings of the various designs.

### Cessna Citation $\mu$-Design

In chapter 2 we performed several robust control designs to achieve desired handling qualities in the pitch axis of the Cessna Citation 500. It was shown how to setup, translate and analyze the robust handling quality problem within the $\mu$ framework. Special attention was paid in demonstrating how to use both, $\mu$ analysis results and $\mu$-syntesis for the determination of balanced trade-off levels between the performance and stability robustness objectives.

- Flexibility in system architecture choice?

  Traditionally the pitch rate command system is obtained in two consecutive design steps. First, the feedback controller is synthesized to provide adequate tracking and robustness properties. Next, a feedforward controller which does not influence the previously obtained robustness properties is developed to shape the command response. This is done by introducing a synthetic short period zero, $T_{\theta 2}$ through command filtering. On the other hand the generalized framework allows us indeed to perform the design in the classical fashion but we have seen that it is more efficient to design in one single step a feedback and feedforward controller fulfilling all requirements at once by using the two degrees-of-freedom control structure on wich an ideal reference model is specified. Such a system setup avoids the iterative command shaping steps needed in a traditional design approach resulting in a more efficient design activity.

  Another difficulty in classical control is that system components such as actuators and sensors are often included in a later stage of the design. This necessitates in most cases several redesigns until results become satisfactory. The general framework does not impose such design limitations since we are free to make the architecture as complex as it is needed. Rather than that, the art in robust control is how to keep the system's complexity low in order to develop a (robust) design that is as simple as possible.

- Flexibility in the inclusion of design specifications?

  The robust control framework is versatile enough for a broad inclusion of typical aeronautical control design specifications. Any signal we would like to control can be penalized by weighting functions that reflect the amplitude boundaries within which the worst case system responses must lie. This is in contrast to LQ techniques where the weighting functions do not have a direct meaning for design. The weighting functions in the $H_\infty$ setting reflect physical levels of signal ratios that must be maintained by the controlled system. It has been showm how performance requirements such as tracking error levels, actuator limitations, stall and acceleration limitations can all be included simultaneously into one single overall objective. In that respect the method is general and efficient since it allows a simultaneous treatment of the requirements. The only restriction of the method comes from the nature of the induced norm used which might conservatively overstate the objectives.

The robusntess issues are handled by the explicit inclusion of uncertainty models. The choice in the nature of the uncertainty model is very flexible. This depends on several related factors. First, it depends on how much in-depth system knowledge is available like the excursion of certain system parameters and on the effort necessary to model them. Then, it depends on the stability margins we want to keep in the face the required performance level. Initially it suffices to choose only crude uncertainty models and refine them only by more specialized parametric uncertainty models when the former do not lead to satifactory robust performance levels. In the design example it was shown how to make a consequent trade-off between robustness and performance by efficiently reformulating the posed problem as a robust performance problem for which we can optimize the structured singular value $\mu$.

- Quality of the design solution?

In the Cessna design example it has been shown the quality of the design solution when viewed as robust performance level measured using the structured singular value $\mu$, depends on both, the choice of the uncertainty models and the performance filters. In the extreme case, when disregarding the inclusion of any uncertainty requirement, the method returns a design solution for which the performance properties are maximized while lacking of robustness.

Another quality of the controller is its transparancy. The controller can be analyzed in the frequency domain and can be interpreted according to the classical loop-shape requirements. We have shown that for increasing tracking error levels the controller yields integral action accordingly.

As regards the implementability of the controller, the method delivers solutions with order equal to the dimension of the original interconnetction structure. This is in general too high for direct implementation. One has to resort for post design controller reduction techniques which form an important part of the design. Especially, the choice of dynamic $D$-scales in the $\mu$-synthesis step has a dramatic influence on the controller order. A clear drawback of the controller reduction phase, is that the designer has no direct control over the preservation of the robust performance property achieved in the synthesis phase. In that respect we should look for ways to include the controller reduction step as an integral part of the design. A practical way to account for the performance degradation due to model reduction is to specify in the controller synthesis phase explicit controller uncertainties. The resulting closed loop system will inherit desired robustness properties at controller level. In this way one can control the loss of the overall robust performance by keeping track of the structured singular value $\mu$ as a function of the system state dimension.

We have also observed that controller synthesis leads at optimality to unstable controllers. We have experienced that the remedy to prevent controller instability is to stop the design iteration at a suboptimal performance level for which the constraints are numerically well conditioned leading to stable controllers.

Another conservatism of $\mu$-synthesis is that it cannot handle real parametric uncertainties but only complex uncertainties. We have seen that uncertainties that are specified to be real need to be assumed complex during controller synthesis phase. Nevertheless, the amount of conservatism of the complex synthesis can be determined using a complex-real $\mu$ analysis and is often in the order of not more than 10 %.

Finally, time domain criteria such a as peak-to-peak or $H_2$ criteria cannot be handled by the method and must be conservatively approximated by $L_2$-induced norm objectives.

- Redesign in the face of evolving requirements?

  For the Cessna Citation design we have demonstrated how to proceed in the face of evolving design requirements. It has been shown how to tighten performance as well as robustness demands under the overall robust performance objective. This means that a redesign amounts in tuning weighting filters and refining the uncertainty models and levels. In that process the structured singular value analysis results guide the designer in making the right trade-off between these competing requirements.

  What is not transparent and theoretically not well-founded is the chosen order of the $D$-scales. As a general rule we keep the order of the $D$-scales as low as possible while seeking for optimal convergence of the structured singular value.

### $\mu$-synthesis for the RCAM

In chapter 4, we have demonstrated the method on a full nonlinear longitudinal and lateral autopilot system for the RCAM design challenge. It has been shown how to set up a $\mu$-synthesis design achieving robustly the desired features without the tedious design effort of a traditionally designed classical controller or a Total Energy Concept system. The so obtained design meets the robustness and performance requirements for the prescribed automatic approach to land phase.

- Flexibility in system architecture choice?

  For the autopilot design the overall system architecture has been divided in a longitudinal and lateral control system. The longitudinal system architecture consists of an integrated autopilot autothrottle control system. Inner loop pitch rate and outer loop speed and altitude tracking are robustly handled in one step. We use a two degrees-of-freedom control structure with prespecified feedforward control of the ideal model reference signals reflected by handling quality requirements. We have shown how to exploit the tracking performance channels to provide a direct mean to shape the interaction level between the altitude and speed channels.

  The lateral controller is designed sequentially with an inner-loop roll angle and sideslip control system and an outer-loop to control the lateral flight-path error. This two-step approach handles in first instance the multivariable inner-loop interaction problem and secondly the slower trajectory dynamics. A consequence of the so chosen approach resulted in a hierarchical control structure that is less complex than a fully centralized controller.

- Flexibility in the inclusion of design specifications?

  We have shown how the design requirements such as performance, passenger comfort, safety limits and control activity can be translated in the weighting functions corresponding to the problem requirements. The interaction level between the various channels is altered (decoupling) by scaling the input command channels and the output error channels. The performance scalings have been chosen in a such a way that they only affect the off-diagonal terms of the multivariable transfer function generated by the previous signal ratios. Robustness has been included using simple lumped dynamics to cover unmodelled higher frequency dynamics.

- Quality of the design solution?

  The choice of the weighting functions has captured most of the performance requirements which have been assessed using Monte-Carlo simulations. An extended parametric uncertainty model to reflect the actual mass, centre of gravity and computational time delay variations subjected to a real $\mu$-analysis has revealed that the chosen lumped uncertainty model for design satisfactorirly captures and achieves the design requirements.

  As for all $\mu$-synthesis designs the order of the controller is high and has been reduced using balanced model reduction techniques. The controller structure itself is transparent and its properties has a straightforward interpretation in the frequency domain.

- Redesign in the face of evolving requirements?

  The same comments concerning the performance-robustness trade-off hold as for the previous design. Using the extended parametric uncertainty model used for the system analysis permits to conduct a refined design to improve the attenuation level in the normal acceleration signal.

**Robust LPV Synthesis for the Missile**

The second part of this thesis was oriented towards the application of various LPV approaches. In chapter 5 we have exhibited LPV designs on a missile problem to benchmark various LPV techniques with unbounded as well as with bounded paramater rates.

- Flexibility in system architecture choice?

  The flexibility in the choice of the system architecture for LPV synthesis is the same as for $\mu$-synthesis. For the purpose of an uniform comparison of the various existing LPV approaches we opted to take the given two degrees-of-freedom structure from literature.

- Flexibility in the inclusion of design specifications?

  The LPV framework is an extension of the $\mu$-framework to parameter varying systems and therefore has an increased flexibility in the inclusion of the design specifications. This greater freedom with respect to the LTI framework is due to the fact that the designer is not only confined to develop gain scheduled control solutions addressing only the parameter dependency of the plant but he has also the extra degree of freedom freedom to seek for gain scheduled solutions for parameter dependent performance requirements. The approach allows the designer to reat the performance channels in the same fashion as the parameter dependent plant by including parameter dependent weighting functions. In the LPV-LFT design approach we only require to develop an LFT representation of the parameter varying part of the system for which we assume the parameters are measurable. The LPV-LFT design approaches all assume infinitely fast parameter variations which is an overly conservative assumption. Dynamic filtering of the parameter measurements could be a way to include parameter rate constraints into the problem formulation. To overcome the conservatism of the LPV-LFT method we also have considered the LPV method with bounded rates. The rate of the parameter variation is directly a design parameter making this design approach even less conservative. Further, this latter method does not impose an LFT representation of the parameter dependent system for control. It suffices to have a set of

LTI systems representing a family of frozen parameter conditions along the system trajectory. We used parameter dependent Lyapunov basis functions in order to generate complex nonlinear scheduling mechanisms. A drawback is, that there is no formal theory guiding the designer in such a choice. Finally, the LPV methods as presented, all allow the inclusion of robusntess issues through appropriate choice of the scaling matrices related to the particular nature and structure of the uncertainty model.

- Quality of the design solution?

  The quality of the LPV design solution is measured in terms of the $L_2$-induced norm of the closed loop parameter dependent system. As expected the method with bounded rates outperforms the LPV-LFT methods at the cost of computational power. Both methods result in controllers having the state dimension of the initial interconnection structure.

- Redesign in the face of evolving requirements?

  Evolving design requirements demand a similar effort as in the case of $\mu$-designs. The only difficulty can occur when one wants to generate very accurate parameter dependent LFT systems. For complex parameter dependencies one has to resort to multidimendional numerical interpolation tools rather than approaching the LFT modelling process with analytical methods. There is also a little known about parameter dependent requirements on command signals, tracking errors or any other signals of interest. Nevertheless, it would be in principle, be possible to extract initial LPV requirements from dedicated experiments on existing systems that are known to behave satisfactorily over the considered design envelope.

**Cessna Citation LPV-Design**

Finally, in chapter 6, we revisited the Cessna Citation handling quality problem studied in chapter 2. The design objective has been refined by seeking a control solution that robustly meets parameter dependent CAP requirements over a large range of the flight envelope.

- Flexibility in system architecture choice?

  The same flexibility holds as in the other cases.

- Flexibility in the inclusion of design specifications?

  The design demonstrated the extended flexibility of the LPV methods in that they not only provide automatically a gain scheduled controller to fixed requirements but also to parameter dependent requirements. The extra design effort was spent in the development of a parameter dependent aircraft model in LFT form for which numerical tools have been used. Further, a parameter dependent ideal model for the CAP requirements has been formulated. Robustness issues have been taken into account using dynamic scalings on the uncertainty channels.

- Quality of the design solution?

  The quality of the design solution is that we have obtained a gain scheduled controller with garanteed levels of performance and robusntess measured in the $L_2$-induced norm. The quality of the solution is comparable to the results obtained using $\mu$-synthesis but now achieved globally over the entire parameter range and over the entire set of parameter dependent requirements. Conservatism is introduced by the assumption of infinite rate of the parameter variation.

- Redesign in the face of evolving requirements?

  The redesign effort is comparable to $\mu$-synthesis.

## 7.2 Some General Conclusions

In the chosen methods we have been guided by many practical considerations that form the criteria for industrial acceptance. From our experience we may claim that the proposed methods establish a natural link between the seemingly theoretical framework and common practice. The proposed methodologies adequately answer needs of the flight control industry but raise even more new issues before we might label the methodology as a mature framework for design.

We were guided by the fact that working in the frequency domain is a prerequisite for the establishment of a bridge between the classical and the robust control school. Notice that this might be misleading and not appropriate in the context of LPV control. This means that notions as the system bandwidth, the cut-off and resonnance frequency that belong to both of the worlds establish a common language for system analysis and communication between the classical and the modern control school. The frequency domain allows also the continuation of well known design rules on loop gains within a mathematically well elaborated framework. Robustness isssues traditionally abstracted using the gain and phase margin concepts are not abondoned and remain conceptually the same but now with extended features. In the Bode plot either destabilizing gain or phase variations can be visuliazed while the Nichols or Nyquist plot permit to visualize simultaneous phase and magnitude shifts. Those variations are assumed to be lumped neglected dynamics and can therefore be represented as complex valued transfer functions.

The structured singular value has been specially developed to generalize these classical loop margins to multiple simultaneous variations. These multiple variations are represented by diagonally augmented uncertainty matrices $\Delta$. The uncertainties themselves are abstracted as norm bounded operators with $||\Delta||_\infty \leq 1$. According to the design phase the uncertainties can be more or less accurately modelled. Nevertheless, the combination of single LTI systems with norm bounded uncertainties is what led to the concept of system sets abstracted through LFTs. The smallest norm bounded block diagonal perturbation destabilizing the LFT system is exactly what defines the structured singular value $\mu$. The structured singular value determines the worst case uncertainty combination making the determinant of $I - \Delta M$ singular. The notion of LFTs plays also an important role in the LPV system theory. The LFT representation allows to generalize the concept of linear time invariant uncertain systems to parameter or time varying uncertain systems.

Such a system formulation together with a suitable gain scheduling synthesis procedure as sketched in the previous chapters increases dramatically the design efficiency. Rather, than performing many single point designs we have at our disposition formal methods that deliver in one design step control solutions with guaranteed levels of robustness and performance.

The evolutionary character of the robust control framework is demonstrated by applying $\mu$-synthesis in combination with the LPV-LFT design method with diagonal scales. The relevance of the sketched procedure should be placed in the light of a more efficient and more powerfull flight control design process.

The overal guidelines when performing such a design is to gradually increase the complexity of the problem. The first major design step is to solve the single operating point robust LTI design. Using $\mu$ synthesis we may determine the optimal compensator in the face of worst case structured uncertainties. The design process involves a plant model with uncertainties made explicit and design weights to reflect the performance objectives. All together the plant model, uncertainties and weighting functions are gathered into the generalized system interconnection for synthesis $P$. The signals affecting the system $P$ are the measurement and control signals, the performance and external input signals and finally the signals that are connected to the uncertainty. The synthesis problem is to find a stabilizing controller $K$ minimizing the closed loop system $M = F_l(P, K)$ in the induced 2-norm. In the $\mu$ analysis framework we require $\mu(M_{11}) < 1$ for robust stability, $\|M_{22}\|_\infty < 1$ for nominal performance and finally $\mu(M) < 1$ to achieve robust performance.

During the control synthesis we have the flexibility to choose the general interconnection system $P$ as complex as we wish. The choice for a well elaborated $P$ might be necessary when we are asked to fulfill simultaneously many requirements. It has to be remarked that the difficulty of the method comes from the availability of mathematically exact formulated requirements. These have to be translated into frequency dependent weighting functions. The crucial step is to find the right translation of the design requirements into a suitable weighting function for design.

The most efficient way to start a design is to disregard the robustness aspect and concentrate at finding and fulfilling the driving performance objectives, such as the tracking, actuator effort and noise requirements. In many cases the designers

enters a long tuning process for weigthing function selection as well as the system scalings for plant conditioning. The first aim is to achieve at each flight condition uniformingly over the whole frequency range a prespecified nominal performance level. It is important to understand how every single requirement influences the rest of the system performance. We recommend to always write down the litteral expression for the generalized system for performance analysis $M_{22}$ and plot all its elements during the design. In most cases we aim at achieving a performance level about $||M_{22}||_\infty \leq 0.5$.
We may address the robustness issues only once the trade-off between the competing nominal performance requirements is sorted out. For robustness considerations it is practical and more efficient to start with crude uncertainty models such as the lumped dynamical models reflecting higher frequency unmodeled dynamics at the actuators and/or at sensors. These are mainly used to impose higher frequency control roll-off. In this way it is possible to meet the required levels of gain and phase margin, mostly chosen to be 6 dB and 45 degrees. In terms of robust stability we aim at attaining a level of $||M_{11}||_\infty \leq 0.5$.

The overall design activity boils down in finding the weighting functions that lead to a balanced trade-off between the required stability robustness and performance properties. In this way we aim at obtaining a closed loop system that performs robustly, i.e., $\mu(M) \leq 1$. In the multivariable case we have seen that achieving robust performance is even more difficult to achieve since $\mu(M)$ is also proportional to the square root of the condition number $\kappa(G)$ of the plant $G$ under consideration. We refer here to the robustness analysis for the longitudinal RCAM controller conducted in chapter 4.

We have seen that by choosing for simplicity $W_{in}W_{id} = I$, the achievable robust performance level is proportional to the square root of the condition number of the plant $(\kappa(G))$. By properly choosing $W_{in}$, we can improve the robust performance level and provide for physical system conditioning. However, by scaling we cannot bring the condition number arbitrarily close to one. Only plant inversion would lead to complete decoupling of the multivariable system and is known to make the resulting system even more susceptible to model uncertainties. This means that we have to accept the limitation imposed by multivariable control. This leads to an achievable level of robust stability and nominal performance imposed by the contributions $||W_{pert}T_i||_\infty$ and $||W_{out}W_pS_oW_{id}W_{in}||_\infty$ that are more conservative than the ones attainable in the single loop case.

The robustness of the RCAM $\mu$-design has been evaluated using worst case $\mu$-analysis. Therefore, the parametric uncertainties given in the requirements have been remodelled in an LFT fashion. These reflect the expected system changes to varying mass, inertia, and centre of graviy location. Also the effect of the computational time delays has been taken into account. The LFT modelling effort has been carried out to compare uniformingly the robustness properties of all RCAM design entries. The tedious LFT model generation has been reported in [MBT97, LMVB96, VLMG96]. The Linear Fractional Transformation describing the parametric uncertainties for the RCAM has been obtained numerically using the PUM toolbox for MATLAB for which the procedure is outlined in the paper given by Lambrechts *etal.* [LTBS93]. It has been shown that the crude lumped uncertainty model used in this thesis to design the $\mu$-controller has efficiently captured the parametric robustness requirements given in section A.2. The robust stability level calculated with the extended LFT model amounts to $\mu(M_{11}) = 0.49$. This means that a gain variation of magnitude $1/0.49 \approx 2$ will destabilize the closed loop system. This corresponds to the usual required 6 dB margin.

In chapter 5 we have extended the $\mu$-synthesis technique to Linear Parameter Varying systems with guaranteed stability and performance levels while taking into account the boundedness of the parameters. Similar to the dynamic $D$-scales applied in $\mu$-synthesis the robustness for LPV systems against uncertainties has been taken into account via the use of constant scalings. The infinite dimensional problem becomes a solvable finite dimensional problem by gridding the parameter space. In this way the synthesis problem is reduced to solving simultaneously finitely many linear matrix inequalities. Though the reformulated problem is computationally tractable it remains for large order problems a computationally very expensive procedure. Solutions are obtained in a matter of days rather than minutes making the design progress inefficiently slow. The gain scheduling feature is obtained by the using parameter dependent Lyapunov functions as basis functions. The choice of these basis functions relies more on trial and error than on theoretically well founded guidelines making the design task even more difficult. The method with bounded parameter rates has been benchmarked on a missile problem for which results are extensively reported in literature. The design has been compared to a series of existing LPV techniques with bounded and unbounded parameter rates and shows to be less conservative than all the methods described in the literature. Implementability of the obtained control solution has been demonstrated through nonlinear simulations showing that the proposed method is successful in achieving the desired performance and robustness goals.

The computational burden imposed by the gridding technique can be overcome by using the more conservative LPV-LFT technique with unbounded parameter rates. Robustness properties have been incorporated into the original LPV-LFT method by again using constant scalings on the devised uncertainty channels. In chapter 6 we have applied the method as outlined in [SBB96] to a handling quality problem for the Cessna Citation 500 aircraft. It has been shown how to incorporate in the design flight condition dependent Gibson's control anticipation requirements. An LFT-LPV model for the Cessna Citation 500 dynamics parametrized in altitude and airspeed has been derived using the PUM toolbox for MATLAB [LTBS93, TL93]. At each local flight condition a $\mu$-synthesis design has been performed to obtain the weighting functions reflecting the desired local performance properties. Once satisfactory weights are obtained from the $\mu$-designs these are incorporated into the robust LPV-LFT synthesis method.

The resulting controller is a parameter dependent LFT controller achieving globally over the flight envelope guaranteed stability robustness and parameter dependent handling qualities requirements. The conservativness imposed by the unboundness of the parameter rates can be overcome by applying in a final design the less conservative robust LPV method with bounded parameter rates as proposed in chapter 5.

## 7.3 Some Remarks

There is a lack of a theory guiding the designer in the choice of the best control architecture. Until now we have only explored centralized control solutions, it would be benefical to reduce the complexity of multivariable centralized control by addressing the problem of finding decentralized robust LPV control solutions.

Concerning the LFT and LPV model generation, there should be a link to practical identification schemes with suitable model validation procedures. Further model reduction schemes for LFT systems should be formally included as new design tools.

Research should also be directed towards the quantitative identification and determination of design specifications. In our case this should be done at moving base flight simulation level to establish suitable identification schemes. Finally, the so developed method should be validated by means of flight tests. Also there should be a common way of system modelling over the various disciplines that are involved in the design of flight control systems.

Model reduction techniques should be incorporated as an integral part of the design rather than a posteriori. This would allow to preserve robust performance in the face of potential model reduction.

LMI solvers although readily available have yet not achieved the maturity to cope in a reasonable computational time for problems with more than twenty states. In that respect solution obtained using the traditional Riccati and Lyapunov equations are much more efficient and numerically better conditioned than LMI solutions.

It is worth to analyze the fragility of robustly designed controllers in the face of uncertainties that daily occur in life. It easy to see that an unexpected software bug still can destroy a robust controller. Maybe a less robust and less performing classical controller would still perform well. How does complexity influence the fragility of an obtained robustness property. It seems that worst case analysis should be elaborated in the much wider context of nonlinear system theory.

The Monte Carlo simulation procedure to establish the robusness properties via time simulations should be formalized with mathematical tools that guarantee these properties. It would be worth to look for formal tools supporting robust simulation techniques that via optimization procedures detect and simulate only worst case system trajectories.

# Bibliography


[AA96] P. Apkarian and R.J. Adams. Advanced gain-scheduling techniques for uncertain systems. *IEEE Transactions on Control System Technology*, 6:21–32, January 1996.

[AGA85] AGARD. Active Control Systems, Review, Evaluations and Projections. *AGARD Conference Proceedings*, CP(384):400, 1985.

[AGA91] AGARD. Handling qualities of unstable highly augmented aircraft. *AGARD Advisory Report*, AR(279):116, 1991.

[Ano75] Anon. Description and flight test results of the NASA F-8 digital fly-by-wire control system. TN D-7843, NASA, February 1975.

[BAG96] J.M. Biannic, P. Apkarian, and W.L. Garrard. Parameter varying control of a high performance aircraft. *IFAC World Congress*, 1996. San Fransisco, USA.

[BAH55] R. Bisplinghoff, H. Ashley, and R. Halfman. *Aeroelasticity*. Addison-Wesley Publishing, Cambridge, MA, 1955.

[BB91] S. Boyd and C. Barratt. *Linear Controller Design: Limits of Performance*. Prentice Hall, 1991.

[BBMF97] F.W. Burcham, J.J. Burken, T.A. Maine, and C.G. Fullerton. Development and flight test of an emergency flight control system using only engine thrust on an MD-11 transport airplane. Technical report, NASA-TP-206217, 1997.

[BDG$^+$93a] G. J. Balas, J. C. Doyle, K. Glover, A. Packard, and R. Smith. *$\mu$-Analysis and Synthesis TOOLBOX, For Use with MATLAB*. The Math Works Inc., first edition, July 1993.

[BDG$^+$93b] G.J. Balas, J.C. Doyle, K. Glover, A. Packard, and R. Smith. *$\mu$-Analysis and Synthesis TOOLBOX, For Use with MATLAB*. The Math Works Inc., first edition, July 1993.

[Bec93] G. Becker. *Quadrtic Stability and Performance of Linear Parameter Dependent Systems*. PhD thesis, University of California at Berkeley, Berkeley, California, December 1993.

[Bec96] G. Becker. Additional results on parameter dependent controllers for LPV systems. *IFAC World Congress*, G:351–356, 1996. San Fransisco, USA.

[BGFB95] S. Boyd, L. El Ghaoui, E. Feron, and V. Balakrishnan. *Linear Matrix Inequalities in System and Control Theory*, volume 15 of *Studies in Applied Mathematics*. SIAM, Philadelphia, 1995.

[BL96] S. Bennani and G. Looye. RCAM design challenge presentation document: the $\mu$-synthesis approach. Technical Report TP-088-11, Group for Aeronautical Research and Technology in Europe (GARTEUR), June 1996. RCAM entry of Delft University of Technology.

[BP92] G.J. Balas and A. Packard. Time-varying controllers for missile autopilots. *1-st IEEE Conference on Control Applications, Dayton,OH*, September 1992.

[BP94] G. Becker and A. Packard. Robust performance of linear parametrically varying systems using parametrically-dependent linear feedback. *System & Control Letters*, 23:205–215, 1994.

[Bro94] R. Brockhaus. *Flugregelung*. Springer-Verlag, Berlin, 1994.

[Bru94] R. W. Brumbaugh. Aircraft model for the AIAA controls design challenge. *Journal of Guidance, Control and Dynamics*, 17, 1994.

[Buf97] J.M. Buffington. Tailless aircraft control allocation. *Proc. of AIAA GNC Conf., New Orleans, LA*, (AIAA-97-3605), Aug 1997.

[BW95] S. Boyd and S. Wu. SDPSOL - A parser/solver semidefinite programs with matrix structure, User's Guide, version alpha, Stanford university. 1995.

[BW98] J.S. Brinker and K.A. Wise. Reconfigurable flight control for a tailless advanced fighter aircraft. *Proc. of AIAA GNC Conf., Boston, MA*, (AIAA-98-4107), Aug 1998.

[BWS98] S. Bennani, D.M.C. Willemsen, and C.W. Scherer. Robust control of linear parametrically varying systems with bounded rates. *Journal of Guidance, Control, and Dynamics*, 21(1), 1998.

[CH69] G.E. Cooper and R.P. Harper. The use of pilot rating in the evaluation of aircraft handling qualities. TN-D 5153, NASA, April 1969.

[Dal85] E. Daley. An update of the fly by wire jaguar equipped with a full-time digital flight control system. *AGARD Conference Proceedings on Active Control Systems, Review, Evaluations and Projections*, CP(384), 1985.

[DB96] W.C. Durham and K.A. Bordignon. Multiple control effector rate limiting. *AIAA Journal of Guidance, Control, and Dynamics*, 19(1):30–37, Jan-Feb 1996.

[DCF+84] J.C. Doyle, C.C. Chu, B. Francis, P. Khargonekar, and G. Stein. Advances in multivariable control. *Lecture Notes* ONR Honeywell Workshop, September 1984.

[DFT92] J.C. Doyle, B.A. Francis, and A.R. Tannenbaum. *Feedback Control Theory*. Macmillan Publishing Company, New York, 1992.

[DGKF88] J. Doyle, K. Glover, P.P. Khargonekar, and B. A. Francis. State-space solutions to standard $H_2$ and $H_\infty$ control problems. *Proc. American Control Conference*, pages 1691–1696, 1988.

[DI88] E. Dowell and M. Ilgamov. *Studies in Nonlinear Aeroelasticity*. Springer-Verlag, New York, N Y, 1988.

[Doy78] J.C. Doyle. Guaranteed margins for LQG regulators. *IEEE Transactions on Automatic Control*, 23:756–757, 1978.

[Doy82] J.C. Doyle. Analysis of feedback systems with structured uncertainties. *IEE Proceedings*, 129, Part D:242–250, November 1982.

[DPZ91] J.C. Doyle, A. Packard, and K. Zhou. Review of LFT's, LMI's and $\mu$. *In IEEE Proceedings of the 30th Conference on Decision and Control, Brighton, England*, 2:1227–1232, December 1991.

[DS81] J.C. Doyle and G. Stein. Multivariable feedback design: Concepts for a classical/modern synthesis. *IEEE Transactions on Automatic Control*, 31(1):4–16, 1981.

[DWS82] J. C. Doyle, J. E. Wall, and G. Stein. Performance and robustness analysis for structured uncertainty. *Proceedings IEEE Conference on Decision and Control*, pages 629–636, 1982.

[Eva85] M. R. Evans. Automatic flight control modes for the AFTI/F-111 mission adaptive wing aircraft. *AGARD Conference Proceedings on Active Control Systems, Review, Evaluations and Projections*, CP(384), 1985.

[FL99] L.F Faleiro and A.A Lambregts. Analysis and tuning of a total energy control system control law using eigenstructure assignment. *Elsevier Aerospace Science & Technology, Special issue on Flight Control Law Design*, (3):127–140, 1999.

[Fli96] Flight Mechanics Action Group 08. *Robust Flight Control Design Challenge Problem Formulation and Manual: the Research Civil Aircraft Model (RCAM)*. GARTEUR FM(AG08)/TP-088-3, second edition, January 1996.

[Foe74] H. Foersching. *Grundlagen der Aeroelastik*. Springer-Verlag, 1974.

[FPEN86] G.F. Franklin, J.D. Powell, and A. Emami-Naeni. *Feedback Control of Dynamic Systems*. Addison-Wesley, , 1986.

[Fra87] B. Francis. *A Course in $H_\infty$ Control Theory*, volume 88. Springer-Verlag, 1987.

[GA93] P. Gahinet and P. Apkarian. An lmi-based parametrization of all $H_\infty$ controllers with applications. *In IEEE Proceedings of the 32nd Conference on Decision and Control, San Antonio, Texas*, 1:656–661, 1993.

[GA94] P. Gahinet and P. Apkarian. A linear matrix inequality approach to $H_\infty$ control. *International Journal of Robust and Nonlinear Control*, 4:421–448, 1994.

[GA95] P. Gahinet and P. Apkarian. A convex characterization of gain-scheduled $H_\infty$-controllers. *IEEE Transactions on Automatic Control*, 40:853–864, May 1995.

[GD88] K. Glover and J.C. Doyle. State-Space Formulae for all Stabilizing Controllers that satisfy an $H_\infty$ Norm Bound and Relations to Risk Sensitivity. *Systems & Control Letters*, 11:167–172, 1988.

[GDN95] L. El Ghaoui, F. Delebecque, and R. Nikoukhah. Lmitool: A user friendly interface for LMI optimization. *User's Guide, Beta version*, 1995.

[GG98] P.J. Goddard and K. Glover. Performance-preserving controller approximation. *IEEE Transactions on Automatic Control*, 43(7), July 1998.

[Gib82] J. C. Gibson. Criteria for handling qualities of military aircraft. *AGARD Conference Proceedings*, CP(333), 1982.

[Gib99] J.C. Gibson. *Development of a Methodology for Excellence in Handling Qualities Design for Fly by Wire Aircraft*. PhD thesis, Delft University of Technology, Faculty of Aerospace Engineering, The Netherlands, 1999.

[GNLC95] P. Gahinet, A. Nemirovski, A.J. Laub, and M. Chilali. *LMI Control Toolbox*. The MathWorks Inc., 1995.

[Gra85] U.P. Graeber. Realisation of relaxed stability on a commercial aircraft. *AGARD Conference Proceedings on Active Control Systems, Review, Evaluations and Projections*, CP(384), 1985.

[Hel95] A. Helmersson. *Methods for robust gain-scheduling*. PhD thesis, Linköping University, Sweden, 1995.

[HJK85] M. J. Walker H. J. Kaul, F. Sella. Flight control system for the experimental aircraft programme demonstrator aircraft. *AGARD Conference Proceedings on Active Control Systems, Review, Evaluations and Projections*, CP(384), 1985.

[HMA+82] R.E. Hoh, D.G. Mitchell, I.L. Ashkenas, R.H. Klein, R.K. Heffley, and J. Hodgkinson. Proposed mil standard and handbook - flying qualities of air vehicles. Volume II: Proposed MIL Handbook AFWAL-TR-82-3081, 1982.

[Hyd95] R.A. Hyde. *$H_\infty$ Aerospace Control Design. A VSTOL Flight Application.* Advances in Industrial Control Series. Springer-Verlag, 1995.

[JR85] C.S. Willey J.J. Rising, W.J. Davis. Demonstration of relaxed stability on a commercial aircraft. *AGARD Conference Proceedings on Active Control Systems, Review, Evaluations and Projections*, CP(384), 1985.

[KBCE90] I. Kaminer, R.A. Benson, E. Coleman, and Y. Ebrahimi. Design of integrated pitch axis for autopilot/autothrottle and integrated lateral axis autopilot/yaw damper for NASA TSRV airplane using integral LQG methodology. NASA-CR 4268, National Aeronautics and Space Administration, Langley Research Center, 1990.

[Keh95] M. Kehoe. A historical overview of flight flutter testing. *Proceedings of the $80^{th}$ AGARD Structures and Materials Panel*, AGARD-CP(566):1–15, 1995.

[Kor85] U. Korte. Some flight test results with redundant digital flight control systems. *AGARD Conference Proceedings on Active Control Systems, Review, Evaluations and Projections*, CP(384), 1985.

[KPR88] P. Khargonekar, I. Peterson, and M. Rotea. $H_\infty$-optimal control with state-feedback. AC(33):786–788, 1988.

[Lam83a] A A. Lambregts. Functional integration of vertical flight path and speed control using energy principles. *Boeing, Seattle*, 1983.

[Lam83b] A A. Lambregts. Integrated system design for flight and propulsion control using total energy principles. *AIAA Paper 83-2561*, 1983.

[Lam83c] A A. Lambregts. Operational aspects of the integrated vertical flight path and speed control system. *AIAA Paper 83-1420*, 1983.

[Lam83d] A A. Lambregts. Vertical flight path and speed control autopilot design using total energy principles. *AIAA Paper 83-2239*, 1983.

[LB96] G. Looye and S. Bennani. Description and analysis of the research civil aircraft model (rcam). Technical report, DUT, Aerospace Engineering, sect. Stability and Control, Delft, June 1996.

[Lee97] L.H. Lee. *Identification and Robust Control of Linear Parameter-Varying Systems.* Phd thesis, University of California at Berkeley, Berkeley, USA, 1997.

[LKS78] R. E. Lee, J.R. Kreitner, and S. A. Stoddart. Flying qualities evaluation of the YC-14 advanced medium STOL transport. TR 77-35, Jan 1978.

[LMVB96] G. Looye, G. Moormann, A. Varga, and S. Bennani. Post-Design Stability Robustness Assessment of the RCAM Controller Design Entries. Technical Report TP-088-35, Group for Aeronautical Research and Technology in Europe GARTEUR-FM(AG08), 1996.

[LS97] L.H. Lee and M. Spillman. Robust, reduced-order, linear parameter-varying flight control for an F-16. *Proceedings of the AIAA Conference on Guidance, Navigation and Control, AIAA-97-3637, New Orleans*, 1997.

[LTBS93] P.F. Lambrechts, J.C. Terlouw, S. Bennani, and M. Steinbuch. Parametric uncertainty modeling using LFTs. *Proceedings of the American Control Conference*, 1993.

[LZD91] W. Lu, K. Zhou, and J. Doyle. Stabilization of LFT systems. *In IEEE Proceedings of the 30-th Conference on Decision and Control, Brighton, England*, 2:1239–1244, 1991.

[Mac89] J.M. Maciejowski. *Multivariable Feedback Design.* Addison Wesley, 1989.

[MBT97] J.-F. Magni, S. Bennani, and J.C. Terlouw. *Robust Flight Control: A Design Challenge.* Lecture Notes in Control and Information Sciences. Springer-Verlag, London, 1997.

[MIL80] MILSPEC2. *Military Specification, Flying Qualities of Piloted Airplanes*, 1980. MIL-F-8785 C.

[MIL87] MILSTD. *Military Standard, Flying Qualities of Piloted Aircraft*, March 1987. MIL-STD-1797.

[Mor85] W. A. Moran. Operational and developmental experience with the F/A-18A digital flight control system. *AGARD Conference Proceedings on Active Control Systems, Review, Evaluations and Projections*, CP(384), 1985.

[MOS95] I. Masubuchi, A. Ohara, and N. Suda. LMI-based controller synthesis: A unified formulation and solution. *Proc. ACC*, 1995.

[MR97] A. Megretski and A. Rantzer. System analysis via integral quadratic constraints. *IEEE Transactions on Automatic Control*, 42:819–830, 1997.

[MS85] D.J. Moorhouse and D.R. Selegan. The STOL and maneuver technology program integrated control system development. *AGARD Conference Proceedings on Active Control Systems, Review, Evaluations and Projections*, CP(384), 1985.

[MTB97] J.F. Magni, J.C. Terlouw, and S. Bennani, editors. *Robust Flight Contol*, chapter RCAM, A $\mu$-Synthesis Approach (I). Springer Verlag, 1997.

[NN93] Yu. Nesterov and A. Nemirovsky. Interior point polynomial methods in convex programming: Theory and application. *SIAM*, 1993.

[NRR93] R.A. Nichols, R.T. Reichert, and W.J. Rugh. Gain scheduling for $H_\infty$ controllers: A flight control example. *IEEE Transactions on Control Systems Technology*, 1(2):69–78, June 1993.

[NSB96] R.G.E. Njio, C.W. Scherer, and S. Bennani. Parametrically varying flight control system design with full-block scalings. *Selected Topics in Identification, Modelling and Control*, 9, 1996.

[Pac94] A. Packard. Gain scheduling via linear fractional transformations. *Systems and Control Letters*, 22(2):79–92, October 1994.

[Pac95] A. Packard. Model reduction from a $\mu$-perspective. *not available*, 1995.

[PD93] A. Packard and J. Doyle. The complex structured singular value. *Automatica*, 29(1):71–109, 1993.

[PDB93] A. Packard, J. Doyle, and G. Balas. Linear, multivariable robust control with a $\mu$ perspective. *American Society of Mechanical Engineering*, 115:426–437, June 1993.

[Pel85] M. Pelegrin. Active control technology: Past, present, future. *AGARD Conference Proceedings on Active Control Systems, Review, Evaluations and Projections*, CP(384), 1985.

[PZP+92] A. Packard, K. Zhou, P. Pandey, J. Leonhardson, and G. Balas. Optimal constant I/0 similarity scaling for full-information and state-feedback control problems. *Systems and Control Letters*, 19(4):271–280, October 1992.

[PZPB91] A. Packard, K. Zhou, P. Pandey, and G. Becker. A collection of robust control problems leading to LMI's. *Proc. IEEE Conf. Decision and Control*, 2:1245–1250, December 1991. Brighton, UK.

[Red60] R. Redheffer. On a certain linear fractional transformation. *J. Math. Phys.*, 39:269–286, 1960.

[RHF75] K.L. Roger, G.E. Hodges, and L. Felt. Active flutter suppression, a flight test demonstration. *Journal of Aircraft*, 12(6):551–557, June 1975.

[RJM85] D. Mc. Ruer, D. Johnston, and T. Meyers. A perspective on superaugmented flight control: Advantages and problems. *AGARD Conference Proceedings on Active Control Systems, Review, Evaluations and Projections*, CP(384), 1985.

[RM94] A. Rantzer and A. Megretski. System analysis via integral quadratic constraints. *Proc. IEEE Conf. Decision Contr.*, pages 3062–3067, 1994.

[Rob91] P.A. Robinson. The modelling of turbulence and downbursts for flight simulators. *UTIAS report No.339, University of Toronto*, 1991.

[SA92] J.S. Shamma and M. Athans. Gain-scheduling: Potential hazards and possible remedies. *IEEE Control Systems Magazine*, 12(3):101–106, June 1992.

[Saf82] M. Safonov. Stability margins of diagonally perturbed multivariable feedback systems. *IEE Proceedings Part D*, pages 251–256, November 1982.

[SB84] R.C. Schwanz and M. Blair. Dynamic modeling uncertainty affecting control system design. *AIAA Dynamcs Specialists Conference, Palm Springs CA*, (84-1057-CP), May 1984.

[SBB96] M. Spillman, P. Blue, and S. Banda. A robust gain-scheduling example using linear parameter varying feedback. *IFAC World Congress*, 1996. San Fransisco, USA.

[Sch90] C. Scherer. *The Riccati Inequality and State-Space $H_\infty$-Optimal Control.* PhD thesis, Universität Würtzburg, Germany, 1990.

[Sch95a] C.W. Scherer. From LMI analysis to multichannel mixed LMI synthesis: A general procedure. *Selected Topics in Identification, Modelling and Control*, 8, 1995.

[Sch95b] C.W. Scherer. *Mixed $H_2/H_\infty$ control, Trends in Control - a European Perspective editor A. Isodori.* Springer-Verlag, Rome, Italy, 1995.

[Sch97] C.W. Scherer. Linear parametrically varying control with full-block scalings. *Proc. IEEE Conf. Decision and Control*, December 1997.

[SD91] G. Stein and J.C. Doyle. Beyond singular values and loop shapes. *Journal of Guidance Control and Dynamics*, 14:5–16, 1991.

[SG95] G. Scorletti and L. El Ghaoui. Improved linear matrix inequality conditions for gain-scheduling. *Proc. 34th IEEE Conf. on Decision and Control*, pages 3626–3631, December 1995.

[SGC97] C.W. Scherer, P. Gahinet, and M. Chilali. Multi-objective output-feedback control. *IEEE Trans. Automat. Control*, 42, 1997.

[SLC89] M. Safonov, D. Limbeer, and R. Chiang. Simplifying the $H_\infty$ theory via loop shifting, matrix pencil and descriptor concepts. *International Journal of Control*, 50:2467–2488, 1989.

[SMN90] M. Sampei, T. Mita, and M. Nakamichi. An algebraic approach to $H_\infty$ output feedback control problems. *Systems and Control Letters*, 14:13–24, 1990.

[SP96] S. Skogestad and I. Postlethwaite. *Multivariable Feedback Control.* John Wiley & Sons, first edition, 1996.

[TL93] J.C. Terlouw and P.F. Lambrechts. A Matlab Toolbox for Parametric Uncertainty Modelling. NLR Technical Publication CR 93455 L, National Aerospace Laboratory NLR, Amsterdam, 1993.

[Tol85] R. D. Toles. Application of AFTI/F-16 task tailored control modes in andvanced multirole fighters. *AGARD Conference Proceedings on Active Control Systems, Review, Evaluations and Projections*, CP(384), 1985.

[Vid78] M. Vidyasagar. *Nonlinear Systems Analysis.* Prentice-Hall, Englewood Cliffs, 1978.

[VLMG96] A. Varga, G. Looye, D. Moormann, and G. Grübel. Automated Generation of LFT-Based Parametric Uncertainty Descriptions from Symbolic Aircraft Models. Technical Report TP-088-36, Group for Aeronautical Research and Technology in Europe GARTEUR-FM(AG08), 1996.

[Wal97] L.A. Walker. Flight testing the X-36, the test pilot's perspective. Technical report, NASA CR-198058, 1997.

[Wea92] K. Wise et al. Linear and nonlinear aircraft flight control for the AIAA control design challenge. *Proceedings of the AIAA Conference on Guidance, Navigation and Control*, 1992.

[Wil96] D.M.C. Willemsen. Robust LPV control for a flight control system. Master's thesis, Delft University of Technology, Department of Aerospace Engineering, Delft, August 1996. masters thesis.

[Wu95] F. Wu. *Control of Linear Parameter Varying Systems.* PhD thesis, University of California at Berkeley, Berkeley, California, June 1995.

[YJ94] P.W. Ying-Jyi. Intelligent control law tuning for AIAA control design challenge. *AIAA Journal of Guidance, Control and Dynamics*, 17, 1994.

[You85] A. D. Young. The aerodynamics of controls. *AGARD Conference Proceedings on Active Control Systems, Review, Evaluations and Projections*, CP(384), 1985. contains also good references there.

[Zam66] G. Zames. On the input-output stability of time-varying nonlinear feedback systems-part 1: Conditions derived using concepts of loop gain, conicity and positivity. *IEEE Transactions on Automatic Control*, AC-11:228–238, April 1966.

[Zam81] G. Zames. Feedback and optimal sensitivity: Model reference transformations, multiplicative seminorms and approximative inverses. *IEEE Transactions on Automatic Control*, 26(2):301–320, February 1981.

[ZDG96] K. Zhou, J.C. Doyle, and K. Glover. *Robust and Optimal Control.* Prentice-Hall, New Jersey, 1996.

[ZK88] K. Zhou and P. Khargonekar. An algebraic riccati equation approach to $H_\infty$ optimization. *Systems and Control Letters*, 11:85–91, 1988.

# Appendix A

# Design Criteria for the RCAM

## A.1 Performance criteria

The performance of control systems are specified in terms of command response characteristics to normalised reference signals, tracking error and disturbance rejection features (see [KBCE90]). Time domain command response characteristics are often defined in terms of linear second order characteristics with rise time $t_r$, settling time $t_s$ and overshoot $M_p$. Rise time is defined as the time the unit step response $y(t)$ takes from $y = 0.10$ to $y = 0.90$, i.e., $t_r = t(y_{90\%}) - t(y_{10\%})$. Settling time is defined as the time for $y(t)$ to achieve 99 percent of its final value. Finally, overshoot is defined as the relative peak of $y(t)$, i.e., $M_p = \frac{(y_{peak} - y(\infty))}{y(\infty)} \times 100\%$ (see [FPEN86]).

- **Lateral deviation.**
  The controlled aircraft's lateral deviation, $e_{yb}(t)$, defined as the difference between the actual and commanded lateral aircraft position, $y(t) - y_c(t)$, should be reduced to 10 percent within 30 $s$.

  There should be very little overshoot in the response to a unit step in lateral command signals at altitudes above 305 $m$ (1000 $ft$), i.e., $M_p < 5\%$. At lower altitudes $M_p$ may increase to 30% in order to obtain higher tracking performance. There should be no steady state error due to constant lateral wind disturbances. In the final phase of flight (landing approach glide path) the lateral deviation from the desired flight path should not exceed that given in figure A.1

- **Altitude response.**
  The controlled system should be able to track altitude commands, $h_c$, with rise time $t_r < 12$ $s$ and settling time $t_s < 45$ $s$. There should be very little overshoot in the response to unit steps in altitude commands at

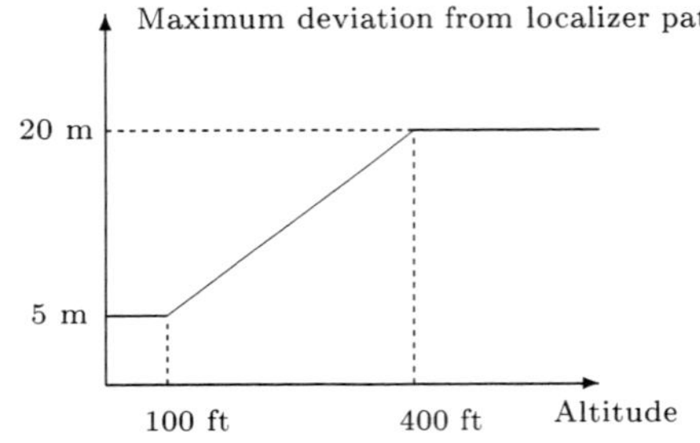


Figure A.1: Maximum lateral deviation

altitudes above 305 *m* (1000 *ft*), i.e., $M_p < 5\%$. At lower altitudes $M_p$ may increase to 30% in order to obtain higher tracking performance. In the final phase of flight (landing approach glide path) the vertical deviation from the desired flight path should not exceed that given in figure A.2

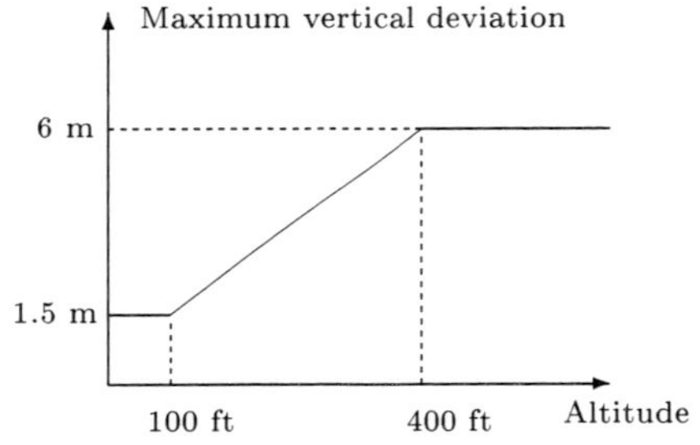


Figure A.2: Maximum vertical deviation

- **Heading angle response.**
The commanded heading angle, $\psi_c$, should be tracked by the actual heading angle, $\psi$, with a rise time $t_r < 10$ *s* and and settling time $t_s < 30$ *s*. There should be very little overshoot in the response to unit steps in heading commands at altitudes above 305 *m* (1000 *ft*), i.e., $M_p < 5\%$. At lower altitudes $M_p$ may increase to 30% in order to obtain higher tracking performance. For unit RMS intensity lateral Dryden gust the RMS of the heading angle error in closed loop should be less than that in open loop.

- **Flight path angle response.**
The commanded flight path angle, $\gamma_c$, should be tracked by the actual flight path angle, $\gamma$, with a rise time $t_r < 5$ *s* and settling time $t_s < 20$ *s*. There should be very little overshoot in the response to unit steps in flight path angle commands at altitudes above 305 *m* (1000 *ft*), i.e., $M_p < 5\%$. At lower altitudes $M_p$ may increase to 30% in order to obtain higher tracking performance.

- **Roll angle response.**

In case of engine failure in still air, the roll angle, $\phi$, should not exceed 10 *deg*; its maximum steady state deviation should not exceed 5 *deg*. During engine failure, sideslip angle $\beta$ should be minimized; the steady state roll angle that is needed to achieve this, should be reduced to zero with an overshoot of less than 50 % when the failed engine is restarted (the failed engine's throttle setting steps back to that of the active engine). Under moderate turbulence conditions $\phi$ should remain smaller than 5 *deg*.

- **Airspeed response.**
  The controlled system's airspeed, $V_A$, should be able to track speed commands, $V_{Ac}$, with a rise time $t_r < 12$ *s* and settling time $t_s < 45$ *s*. There should be very little overshoot in the step response to speed commands at altitudes above 305 *m* (1000 *ft*), i.e., $M_p < 5\%$. At lower altitudes $M_p$ may increase to 30% in order to obtain higher tracking performance. In the presence of a wind step with an amplitude of 13 *m/s* (25 *kts*) there should be no deviation in the airspeed larger than 2.6 *m/s* (5 *kts*) for more than 15 *s*. There should be no steady state error due to constant wind disturbances.

- **Heading rate.**
  In case of engine failure, the maximum heading rate, $\dot{\psi}$, should be less than 3 *deg/sec*.

- **Cross coupling between airspeed $V_A$ and altitude $h$.** For a step in commanded altitude $h_c$ of 30 *m*, the peak value of the transient of the absolute error between $V_A$ and commanded airspeed $V_{Ac}$ should be smaller than 0.5 *m/s* (1 *kt*). Conversely, for a step in commanded airspeed $V_{Ac}$ of 13 *m/s* (25 *kts*), the peak value of the transient of the absolute error between $h$ and $h_c$ should be smaller than 10 *m*.

## A.2 Robustness criteria

- **Centre of gravity variation.**
  Stability and sufficient performance should be maintained for horizontal centre of gravity variations between 15 and 31 % and vertical centre of gravity variations between 0 and 21 % of the mean aerodynamic chord.

- **Mass variations.**
  Stability and sufficient performance should be maintained for aircraft mass variations between 100000 to 150000 *kg*.

- **Time delay.**
  Stability and sufficient performance should be maintained for transport delays from 50 to 100 *ms*.

## A.3 Ride quality criteria

Ride quality criteria should ensure sufficient passenger and pilot comfort. The following specifications are designed to obtain an acceptable level.

- **Maximum vertical acceleration.**
  Under normal conditions during manoeuvres (no turbulence) the vertical acceleration at the centre of gravity should be minimized; it should be less than ± 0.05 *g*[1].

- **Maximum lateral acceleration.**
  Under normal conditions during manoeuvres (no turbulence) the lateral acceleration at the centre of gravity should be minimized; it should be less than ± 0.02 *g*.

- **Damping.**
  Unless stated differently, there should be no overshoot in any step response of any controlled variable at altitudes above 305 *m* (1000 *ft*). Below that altitude overshoot may increase to 30 % in order to obtain higher tracking performance.

## A.4 Safety criteria

- **Airspeed.**
  The airspeed must always be larger than $1.05 \times V_{\text{stall}}$, where $V_{\text{stall}} = 51.8$ *m/s* denotes the stall speed, i.e. the speed below which the aircraft is unable to maintain equilibrium flight.

- **Angle of attack.**
  To the stall speed corresponds an angle of attack $\alpha_{\text{stall}} = 18$ *deg*. For safety reasons a value of 12 *deg* is considered acceptable.

- **Roll angle.**
  The maximum roll angle $\phi$ should be limited to 30 *deg*.

- **Sideslip angle response.**
  At all times, sideslip angle $\beta$ should be minimized. For unit RMS intensity lateral Dryden gust the RMS of the sideslip angle in closed loop should be less than that in open loop.

## A.5 Control activity criteria

- **Actuator effort minimization.**
  Under moderate turbulence conditions, mean actuator rates for aileron, tailplane and rudder should be less than 33 % of the maximum rates.

[1] This value is used in industry during the design phase, in fact the vertical and lateral acceleration limits depend on frequency. They are even lower at 2 *Hz*.

The numerical values for rate limits and saturations are given as follows.

- Rate limits for aileron deflection are: $-25 \leq \dot{\delta_a} \leq 25$ deg/s
  Saturations of aileron deflection are: $-25 \leq \delta_a \leq 25$ deg
- Rate limits for elevator deflection are: $-15 \leq \dot{\delta_e} \leq 15$ deg/s
  Saturations of elevator deflection are: $-25 \leq \delta_e \leq 10$ deg
- Rate limits for rudder deflection are: $-25 \leq \dot{\delta_r} \leq 25$ deg/s
  Saturations of rudder deflection are: $-30 \leq \delta_r \leq 30$ deg

- **Engine effort minimization.**
  Under moderate turbulence conditions, mean throttle rate should be less than 15 % of the maximum rate.

  The numerical values for rate limits and saturations are given as follows.

  - Rate limits for throttle movement are: $-1.6 \leq \dot{\delta}_{th_i} \leq 1.6$ deg/s
    Throttle saturations are: $0.5 \leq \delta_{th_i} \leq 10$ deg

# A.6 Translation of design criteria into evaluation criteria

For the evaluation criteria the same classification is used as was given in the definition of the design criteria.
For each of these indicators and for each of the four trajectory segments a single number is calculated. This number should not be considered to be the final word on overall autopilot performance: it is an indicator for the driving autopilot features such as tracking, disturbance rejection, decoupling and axis coordination. In most cases the measure is chosen such that a value of smaller than one is acceptable.

To evaluate the dynamic behaviour of the autopilot, we will consider several plots of key variables during each of the segments. We will compare the shape of the actual trajectory with the demanded trajectory and provide bounds that should be respected for good performance. Similarly, we will plot the most important deviations from the desired trajectory.

## A.6.1 Segment I

For segment I we will plot a plan view of the reference trajectory and the four trajectories defined in section 4.1, and then superimpose the bounds given in figure A.3.
The points **a** and **b** correspond to the beginning and end of the engine failure segment.

**Performance.**
The bound of 20 $m$ is given to account for the effect of turbulence. During

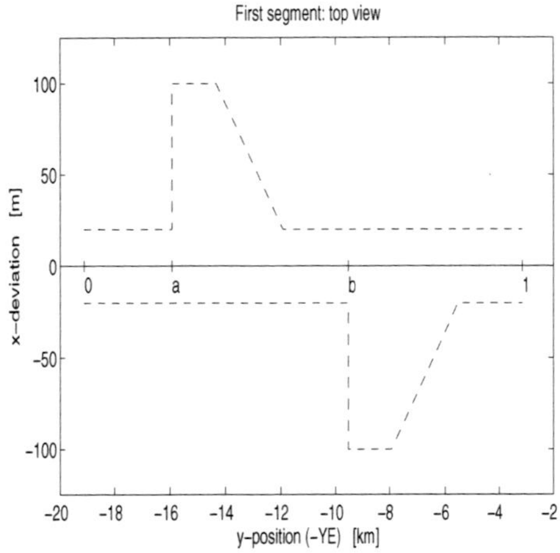


Figure A.3: Segment I: the effect of engine failure with bounds

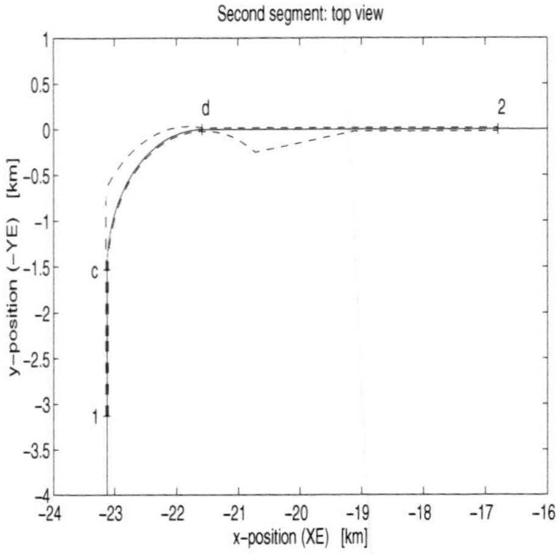


Figure A.4: Segment II: plan view of the 90 degrees turn with bounds

engine failure, we allow a maximum lateral deviation of 100 $m$ that should be quickly reduced to less than 20 $m$ at the end of the segment (when the aircraft should be stabilized again). With $e_{yb}$ denoting the lateral deviation in body co-ordinates for the trajectory with nominal centre of gravity and time delay we will use

$$\left(\max_t \frac{|e_{yb}(t)|}{100} + \frac{|e_{yb}(t_1)|}{20}\right)/2 \tag{A.1}$$

as a measure that should be smaller than one for sufficient performance. Note that during the entire segment the maximum deviation of 100 $m$ should not be exceeded, and that at the end of the segment ($t_1$ corresponds with point **1**) the maximum deviation of 20 $m$ is taken into account.

**Robustness.**
The maximum differences between the lateral deviation of the trajectories with nominal and perturbed centre of gravity and nominal and maximal time delays are considered:

$$\Delta_{eyb}(t) := \max\left(|e_{ybmax}(t) - e_{yb}(t)|, |e_{ybmin}(t) - e_{yb}(t)|\right) \tag{A.2}$$

We will allow differences of 10% of the maximal allowable lateral deviations:

$$\left(\max_t \frac{\Delta_{eyb}(t)}{10} + \frac{\Delta_{eyb}(t_1)}{2}\right)/2 \tag{A.3}$$

should be smaller than one.

**Ride quality.**

The maximum lateral acceleration will be limited by:

$$\frac{|n_y|}{0.2} < 1 \tag{A.4}$$

i.e. $|n_y|$ should be smaller than $0.2g$: under normal flight conditions this value should be much lower ($0.02g$, see section A.3), but engine failure is an emergency situation such that an unusually large lateral acceleration is acceptable.

**Safety.**
During the segment, the maximum angle of attack $\alpha$ will be limited:

$$\max_t \left(\frac{|\alpha(t)|}{12}\right)^3 < 1 \tag{A.5}$$

This implies we accept $\alpha = 12$ *deg*; the power is taken to stress the fact that $\alpha > 12$ *deg* quickly becomes unacceptable (stall situation).

**Control activity.**
The rudder actuator effort will be considered that is needed to stabilize the aircraft after engine failure is lifted: this is calculated as:

$$\int_{t_b}^{t_1} \delta_R^2 dt \tag{A.6}$$

with $t_b$ denoting the end of engine failure (corresponds to point **b** in figure A.3). This value is not 'normalised to one' as it is not clear what bounds can be obtained: it will act as a value for relative comparison of controllers.

## A.6.2 Segment II

For segment II we will plot a plan view of the reference trajectory and the four trajectories defined in section 4.1, and then superimpose the bounds given in figure A.4. Furthermore, to obtain a better insight in the results, we will plot lateral deviations with bounds as given in figure A.5.

**Performance.**
The maximum lateral deviation (due to the turn) and the lateral deviation at the end of the segment (when the aircraft should be stabilized again) are considered:

$$\left(\max_t \frac{|e_{yb}(t)|}{200} + \frac{|e_{yb}(t_2)|}{20}\right)/2 < 1 \tag{A.7}$$

Note that during the entire segment a maximum deviation of 200 $m$ should not be exceeded, and that at the end of the segment ($t_2$ corresponds with point **2**) a maximum deviation of 20 $m$ is taken into account.

**Robustness.**
As in segment I, the maximum differences between the lateral deviation of the trajectories with nominal and perturbed centre of gravity locations and nominal

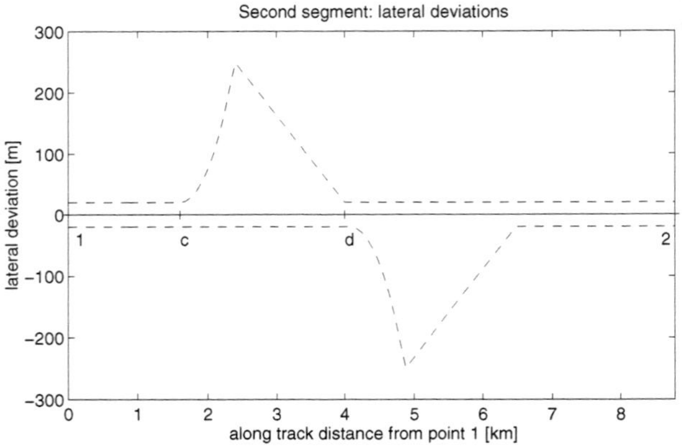


Figure A.5: Segment II: lateral deviations during the 90 degrees turn with bounds

and maximal time delays are considered. Again, we will allow differences of 10% of the maximal allowable lateral deviations:

$$\left(\max_t \frac{\Delta_{eyb}(t)}{20} + \frac{\Delta_{eyb}(t_2)}{2}\right)/2 < 1 \tag{A.8}$$

**Ride quality.**

As in segment I, the maximum lateral acceleration $n_y$ will be considered:

$$\frac{|n_y|}{0.02} < 1 \tag{A.9}$$

i.e. $|n_y|$ should be smaller than 0.02 $g$.

**Safety.**

As in segment I, the maximum angle of attack during the segment is limited:

$$\max_t \left(\frac{|\alpha(t)|}{12}\right)^3 < 1 \tag{A.10}$$

**Control activity.**

The rudder and aileron actuator effort is calculated as:

$$\int_{t_1}^{t_2} (\delta_R^2 + \delta_A^2)\, dt \tag{A.11}$$

This value is not 'normalised to one' as it is not clear what bounds can be obtained: it will act as a value for relative comparison of controllers.

### A.6.3 Segment III

For segment III we will plot a side view of the four trajectories defined in section 4.1. Figure A.6 shows the reference trajectory, the start and end points

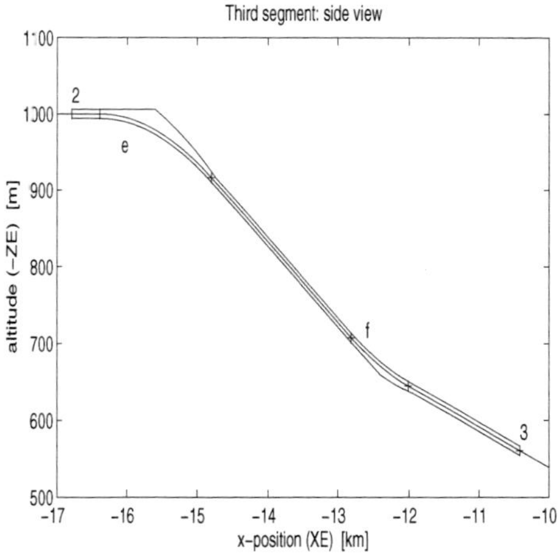


Figure A.6: Segment III: side view of the -6 and -3 degrees glideslope captures with bounds

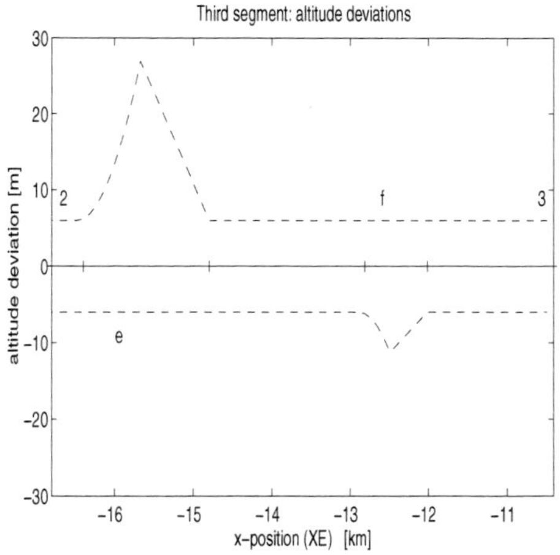


Figure A.7: Segment III: vertical deviations during the -6 and -3 degrees glideslope captures with bounds

of the segment (points **2** and **3**) and the considered bounds; the command actions are labelled with **e** and **f**.
We will also plot the vertical deviation of the trajectories and overlay the bounds shown in figure A.7.

**Performance.**
The maximum vertical deviation during the capture of the -6 degrees glideslope and the vertical deviation at the end of the segment (when the aircraft should be stabilized again) are considered. Furthermore, speed variations should be kept small in spite of the change in required angle of attack. With $e_{zb}$ denoting the vertical deviation in body co-ordinates for the trajectory with nominal centre of gravity and time delay, we will demand

$$\left(\max_t \frac{|e_{zb}(t)|}{20} + \frac{|e_{zb}(t_3)|}{6} + \frac{|V_A - V_{Ac}|}{4}\right)/3 < 1 \tag{A.12}$$

for sufficient performance. Note that during the entire segment a maximum deviation of 20 $m$ should not be exceeded, and that at the end of the segment ($t_3$ corresponds with point **3**) a maximum deviation of 6 $m$ is taken into account. Speed variations should not exceed 4 $m/s$, i.e. 5% of $V_{Ac} = 80$ $m/s$).

**Robustness.**
The maximum differences between the vertical deviation of the trajectories for the nominal and perturbed centre of gravity locations and nominal and maximal time delays are considered:

$$\Delta_{ezb}(t) := \max\left(|e_{zbmax}(t) - e_{zb}(t)|, |e_{zbmin}(t) - e_{zb}(t)|\right) \tag{A.13}$$

We will allow differences of 10% of the maximal allowable vertical deviations:

$$\left(\max_t \frac{\Delta_{ezb}(t)}{2} + \frac{\Delta_{ezb}(t_3)}{0.6}\right)/2 < 1 \tag{A.14}$$

**Ride quality.**

The maximum vertical acceleration $n_z$ will be limited:

$$\frac{|n_z|}{0.1} < 1 \tag{A.15}$$

i.e. $|n_z|$ should be smaller than 0.1 $g$.

**Safety.**
Again, the maximum angle of attack during the segment is limited:

$$\max_t \left(\frac{|\alpha(t)|}{12}\right)^3 < 1 \tag{A.16}$$

**Control activity.**

The tailplane actuator effort is calculated as:

$$\int_{t_2}^{t_3} \delta_T^2 dt \tag{A.17}$$

This value is not 'normalised to one' as it is not clear what bounds can be obtained: it will act as a value for relative comparison of controllers.

### A.6.4 Segment IV

For segment IV we will plot a side view of the four trajectories defined in section 4.1. The wind shear model, the desired trajectory through it, and the bounds are given in figure A.8.
As mentioned before, the wind shear model is a two dimensional model derived from [Rob91]. Along the trajectory, the aircraft will be faced with a headwind going up to about $Wx_E = -7$ $m/s$, then windspeed will change to a tailwind of about $Wx_E = 7$ $m/s$, combined with a downdraught of about $Wz_E = 8$ $m/s$ (see figure A.9). The result of this will be a drastic decrease in aircraft energy: the aircraft will not be able to stay on the desired trajectory. The size of the longitudinal deviation and the time until recovery will be measures for evaluation of the controller.
For this reason we will also plot the longitudinal deviations with bounds as given in figure A.10.

**Performance.**
The maximum longitudinal deviation (due to the wind shear) and the longitudinal deviation at the end of the segment (when the aircraft should be within

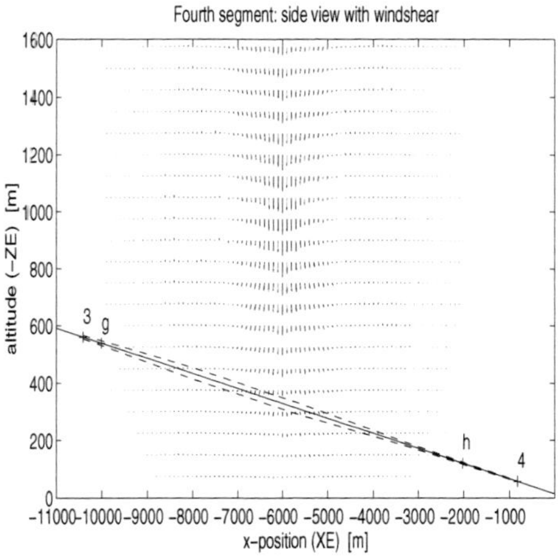


Figure A.8: Segment IV: side view of the final approach with wind shear and bounds

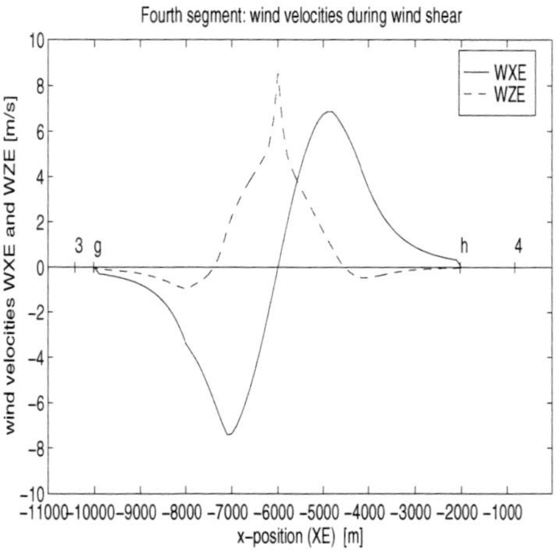


Figure A.9: Segment IV: wind speeds along the trajectory

the decision window) are considered for the trajectory with nominal centre of gravity and time delay:

$$\left(\max_t \frac{|e_{zb}(t)|}{20} + \frac{|e_{zb}(t_4)|}{1.5}\right)/2 < 1 \tag{A.18}$$

Note that during the entire segment a maximum deviation of 20 $m$ should not be exceeded, and that at the end of the segment ($t_4$ corresponds with point **4**) a maximum deviation of 1.5 $m$ is taken into account.

**Robustness.**
As in segment III, the maximum differences between the vertical deviation of the trajectories for the nominal and perturbed centre of gravity locations and nominal and maximal time delays are considered. Again, we will allow differences of 10% of the maximal allowable vertical deviations:

$$\left(\max_t \frac{\Delta_{ezb}(t)}{2} + \frac{\Delta_{ezb}(t_4)}{0.15}\right)/2 < 1 \tag{A.19}$$

**Ride quality.**

As in segment III, the maximum vertical acceleration $n_z$ will be limited:

$$\frac{|n_z|}{0.2} < 1 \tag{A.20}$$

i.e. $|n_z|$ should be smaller than $0.2g$ (in segment III this value is lower, but wind shear is an emergency situation).

**Safety.**

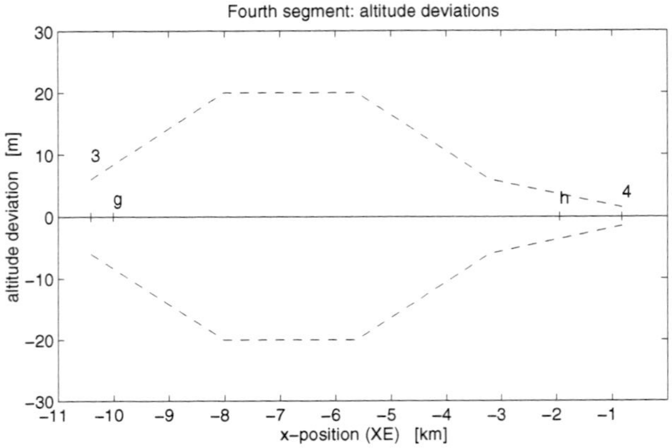


Figure A.10: Segment IV: vertical deviations during the final approach with bounds

We will consider whether the aircraft is within the decision window at the end of the segment. Lateral, vertical and speed variations are limited to 5 $m$, 1.5 $m$ and 3 $m/s$ respectively as follows:

$$\sqrt{\frac{1}{3}\left((\frac{e_{yb}}{5})^2 + (\frac{e_{zb}}{1.5})^2 + (\frac{(V_A - V_{Ac})}{3})^2\right)} < 1 \qquad \text{(A.21)}$$

**Control activity.**

The tailplane and throttle actuators effort is calculated as follows:

$$\int_{t_3}^{t_4} \left(\delta_T^2 + (\delta_{TH_1} + \delta_{TH_2})^2\right) dt \qquad \text{(A.22)}$$

This value is not 'normalised to one' as it is not clear what bounds can be obtained: it will act as a value for relative comparison of controllers.

# Samenvatting en Curriculum Vitae

Dit proefschrift beschrijft de toepassing van regelingen op de beweging van vliegtuigen met behulp van een aantal moderne robuuste regel methodieken die in de laatste tien jaar zijn ontwikkeld binnen het $H_\infty$ raamwerk.

Als eerste passeren de algemene definities die betrekking hebben op het optimaal $H_\infty$-regelen en de $\mu$-synthese de revue.
In het bijzonder, zullen we een algemeen raamwerk voor regelaar synthese en analyse beschouwen dat gebaseerd is op zekere matrix normen en een algemene systeem representatie, de Lineaire Fractionele Transformaties LFT, om de begrippen robuuste stabiliteit, nominale prestatie en robuuste prestatie te definieren.
Vanuit het oogpunt van de regelaar-ontwerper hebben wij aandacht geschonken om de ontwerpstappen binnen het nieuwe raamwerk helder weer te geven. Dit is gedaan aan de hand van de toepassing op een ontwerp van een twee graden van vrijheid "pitch rate" regelsysteem wat in de robuuste zin aan "Handling Quality" eisen moet voldoen.

In een eerste stap, zullen we laten zien op welke wijze simultane prestatie eisen, zoals het volgen van referentie signalen, het onderdrukken van verstoringen en het beperken van regelactiviteit, op een wiskundige wijze geformuleerd moeten worden door geschikte keuzes van weegfuncties.
Ontwerpregels worden gegeven ten aanzien van de keuze van weegfuncties om uitgebalanceerde systeem prestaties af te dwingen. Als de nominale systeemprestaties eenmaal vast liggen worden er verschillende robuustheids eisen aan het probleem toegevoegd.
Gebruik makend van de gestructureerde singuliere waarde als maat voor de robuuste prestaties van systemem wordt geïllustreerd op welke wijze een compromis gemaakt kan worden tussen robuustheids-eisen en prestatie-eisen.
In een tweede stap, om de flexibiliteit te demonstreren van het gestructureerde singuliere waarde concept toegepast op complex multivariabele systemen zoals die optreden in industriele luchtvaarkundige toepassingen, hebben wij een referentie probleem ontwikkeld voor het RCAM (Research Civil Aircraft Model) vliegtuig.

We laten de wijze zien waarop een robuust regelsysteem word ontworpen dat voldoet aan multivariabel volggedrag, veiligheids beperkingen zoals overtrek of rolhoek begrenzingen en stuur coordinaties zoals die benodigd zijn bij zuivere bocht coordinaties en gewenste gladde hoogte en snelheids intercepts.

Bijzondere aandacht is besteedt om een ontwerp te demonstreren van een automatische piloot met een vergelijkbaar gedrag als dat die van een Total Energy Concept Systeem (TECS) met nog verder ingebouwde prestatie robuustheids garanties.

De voordelen van de robuuste aanpak ten opzichte van TECS is dat men rekening kan houden met algemenere prestatie-eisen en tegelijkertijd met meerdere robuustheids-eisen. Aan de andere kant berust het TECS systeem op physische inzichten uit de klassieke vliegmechanica en is de structuur van de regelaar transparant.
In dit opzicht is de TECS regelaar makkelijker te adapteren (schedulen) aan wisselende bedrijfsomstandigheden dan een multivariabele regelaar verkregen met robuuste optimale technieken. Niettemin, is deze adaptie prodedure vliegtuig gebonden en niet algemeen toepasbaar. Ook voldoet het zo verkregen regelsysteem niet aan formele stabiliteits garanties die rekening houden met het in de tijd varierende karakter van het systeem.
Om deze tekortkomingen te ondervangen is er in een tweede deel van dit proefschrift onderzoek gedaan naar wiskundig formele "Gain Scheduling" technieken. We hebben technieken beschouwd die nauwe verwantschap vertonen met de aanpak uit de robuuste lineare systeem theorie, maar nu verder toegesneden op systemen met Linear Varierende Parameters (LPV).
Beiden, zowel de analyse als de synthese resultaten zijn geformuleerd in termen van Lineaire Matrix Ongelijkheden (LMI), die opgelost kunnen worden met geschikte numerieke gereedschappen uit de convexe optimalisatie theorie.

We maken gebruik van Integraal Kwadratische Voorwaarden (IQC) om schalingsmatrices te bepalen voor onzekere LPV systemen. Speciale regelaar parametrisatie formules zijn gebruikt om een grote klasse van LPV synthese technieken te beschrijven. We hebben twee LPV-LFT technieken onderzocht die uitgaan van oneindige snelle parameter variaties en een LPV techniek die rekening houdt met de snelheid van de parameter variaties. Deze methoden zijn geïllustreerd en vergeleken aan de hand van een referentie ontwerp voor een geleide raket en een "Handling Quality" ontwerp probleem voor de Cessna Citation wat voldoet aan het Control Anticipation Parameter criterium van Gibson.

Het laatste hoofdstuk geeft een overzicht weer van de bijdrage van dit proefschrift aan de "flight control system design" gemeenschap. Ook worden er conclusies uit deze studie getrokken en aanbeveling voor verder onderzoek aangedragen.

Uit deze onderzoeksinspanning is gebleken dat de robuuste en LPV concepten zeer geschikt zijn voor de huidige "flight control" ontwerp opgave. Deze nieuwe concepten zijn in geen geval bedoeld om de klassieke ontwerpmethodieken te vervangen. Ze moeten gezien worden als formele generalisaties van de classieke technieken naar multivariabele in de tijd varierende systemen.

# Curriculum Vitae

| | |
|---|---|
| Naam | Samir Bennani |
| 8 mei 1960 | Geboren te Casablanca |
| 1979 | Baccalaureat (Academie de Bordeaux) |
| 1981-1991 | Lucht- en Ruimtevaarttechniek aan de Technische Universiteit Delft |
| 1991-1996 | Promovendus bij de leerstoel Besturing en Simulatie, L&R TU Delft |
| 1996-1997 | Onderzoeker bij Fokker Space, te Leiden |
| 1998- | Universitair docent bij de leerstoel Besturing en Simulatie |